NUREG–1367

Functional Capability of Piping Systems

U.S. Nuclear Regulatory Commission

Office of Nuclear Reactor Regulation

D. Terao, E. C. Rodabaugh

Functional Capability of Piping Systems

Manuscript Completed: October 1992
Date Published: November 1992

D. Terao, E. C. Rodabaugh

Division of Engineering
Office of Nuclear Reactor Regulation
U.S. Nuclear Regulatory Commission
Washington, DC 20555

ABSTRACT

General Design Criterion 1 of Appendix A to Part 50 of Title 10 of the *Code of Federal Regulations* requires, in part, that structures, systems, and components important to safety be designed to withstand the effects of earthquakes without a loss of capability to perform their safety function. The function of a piping system is to convey fluids from one location to another. The functional capability of a piping system might be lost if, for example, the cross-sectional flow area of the pipe were deformed to such an extent that the required flow through the pipe would be restricted.

The objective of this report is to examine the present rules in the American Society of Mechanical Engineers Boiler and Pressure Vessel Code, Section III, and potential changes to these rules, to determine if they are adequate for ensuring the functional capability of safety-related piping systems in nuclear power plants.

CONTENTS

FIGURES

TABLES

NOMENCLATURE

B_1, B_2, B_{2b}, B_{2r} = Code stress indices

D = pipe mean diameter

D_o = pipe outside diameter

Mi = resultant moment, used in Code Equation (9)

M_L = calculated limit moment

Mm = maximum measured moment

P = internal pressure

S = calculated stress based on elastic response spectrum analysis with $+/-15\%$ peak broadening and with either 2% or 5% damping

Sh = Code allowable stress, Class 2 piping

Sm = Code allowable stress intensity, Class 1 piping

Sp = stress due to internal pressure $PD_o/2t$

Sw = stress due to weight

Sy = yield strength of material

t = wall thickness

Z = nominal section modulus of piping component

Other symbols are defined where used in text or tables.

1 INTRODUCTION

General Design Criterion 1 of Appendix A to Part 50 of Title 10 of the *Code of Federal Regulations* (10 CFR) requires that structures, systems, and components important to safety be designed to withstand the effects of earthquakes without a loss of capability to perform their safety function. The function of a piping system is to convey fluids from one location to another. Sizing of the pipe usually involves a compromise as size increases between increasing installed costs and decreasing pressure drop. Functional capability of a piping system might be lost if, for example, displacements were large enough to "crimp" a pipe cross section and thus reduce the flow area.

The Code* does not address the functional capability of piping systems; rather, it addresses pressure boundary integrity. Accordingly, it does not necessarily follow that meeting Code rules will ensure functional capability.

The objective of this report is to examine present Code rules, and potential changes to these rules, to see if they are sufficient to ensure maintenance of functional capability.

---

*"Code" as used in this report refers to the American Society of Mechanical Engineers (ASME) Boiler and Pressure Vessel Code (Ref. 1). Portions of the Code are identified as they appear in the Code (e.g., NB-3652). For the purpose of this report, NC-3600 (Class 2 piping) and ND-3600 (Class 2 piping) are identical; hence, reference is to NC-3600 for Class 2 piping.

2 BACKGROUND

2.1 Present Code Rules

Primary loads, such as internal pressure and weight, in combination with other loads such as those due to earthquakes are controlled* in the Code by Equation (9) in NB–3652 (Class 1 piping) and Equations (8) and (9) in NC–3652 and –3653.1 (Class 2 piping). These Code equations are

$$B_1 PD_O /2t + B_2 Mi/Z < \text{lesser of } X \times Sx \text{ or } Y \times Sy \qquad (1)$$

The symbols are defined in the "Nomenclature" section of this report. The values of X and Y are

Condition	Class 1 Piping X	Y	Class 2 Piping X	Y
Design	1.5	---	1.5	---
Level A	---	---	1.8	1.5
Level B	1.8	1.5	1.8	1.5
Level C	2.25	1.8	2.25	1.8
Level D	3.0	2.0	3.0	2.0

In Equation (1), $Sx = Sm$ (allowable stress intensity) for Class 1 piping, and $Sx = Sh$ (allowable stress) for Class 2 piping. Values of Sm are usually greater than those of Sh. For example, for SA106 Grade B carbon steel at 500°F, $Sm = 18.9$ ksi, while $Sh = 15.0$ ksi. However, for austenitic stainless steels at elevated temperatures, Sm is almost the same as Sh; for example, for SA312 Type 304 stainless steel at 650°F, $Sm = 16.2$ ksi, and $Sh = 15.9$ ksi. The material yield strength, Sy, is 17.9 ksi; thus, $Sm/Sy = Sh/Sy = 0.9$.

It should be emphasized that the resultant moment amplitude, Mi, includes both steady-state loads, such as weight, and dynamic loads, such as those caused by earthquakes. In Level D applications, the dynamic loads have usually been the major contributor to Mi. However, increasing the Level D stress limits is being considered. This possibility, along with the use of higher (e.g., 5%) damping in evaluating the response of piping systems to dynamic loads, makes it more important to recognize that Mi represents combinations of steady-state loads with dynamic loads.

---

*In NB–3658 and NC–3658, rules are given for the analysis of flanged joints. These rules are based on the prevention of excessive leakage at the joints. Because loss of functional capability of a flanged joint (without loss of pressure boundary integrity) is deemed to be incredible, the rules for flanged joints are not considered any further in this report.

2.2 Nuclear Regulatory Commission's Position on Piping Functionality

In the early 1970s, the stress limit of $3Sm$ was considered to be quite high, relative to prior stress limits used in piping design. For example, the industrial piping code, USAS B31.3–1967 (Ref. 2), permitted stresses of $1.2 \times Sh$ for loadings acting not more than 1% of the time. (Earthquake loadings fit in this category.) The concerns of the U.S. Nuclear Regulatory Commission (NRC) related to functional capability of piping with the $3Sm$ limit resulted in the preparation of NUREG/CR–0261 (Ref. 3).

Reference 3 includes summaries of available data on static load capacities of straight pipe, elbows, branch connections, tees, and other piping components. In this reference, several changes in B-indices were suggested:

(1) Restrict application of B-indices to $D_O/t < 50$ (because of the buckling of straight pipes with $D_O/t > 50$).

(2) Decrease B_1 for elbows from 1.0 to 0.5.

(3) Decrease B_2 for elbows from $0.75 \times C_2$ to $0.67 \times C_2$.

(4) Decrease B_{2b} for branch connections from $0.75 \times C_{2b}$ to $0.50 \times C_{2b}$.

However, the data in Reference 3 indicated that Equation (1), $3Sm$ limit, as applied to straight pipe ($B_1 = 0.5$, $B_2 = 1.0$), was the least defensible from the standpoint of static load capacity. For straight pipe, limit load theory (confirmed by cited tests) gives the bending moment, M_L, at zero pressure of

$$M_L = (4/\pi) ZSy \qquad (2)$$

For austenitic stainless steels at elevated temperatures, $Sm = 0.9Sy$. Equation (1) with a $3Sm$ limit would permit application of a moment of $2.7/(4/\pi) = 2.1$ times the static bending limit moment. In Reference 3, it was suggested that the Level D limit be made the lesser of $3Sm$ or $2Sy$.

From the standpoint of functional capability, the $2Sy$ limit is not defensible if Mi in Equation (1) comes from static loads such as weight or steady-state relief valve thrust. Thus, Reference 3 indicated that even the $2Sy$ Level D limit was not clearly defensible for assurance of functional capability.

NRC Standard Review Plan Section 3.9.3 (Ref. 4), Appendix A, states:

2.3 Functional Capability

The design of Class 1, 2, and 3 piping components shall include a functional capability assurance program. This program shall demonstrate that the piping components, as supported, can retain sufficient dimensional stability at service conditions so as not to impair the system's functional capability. The program may be based on tests, analysis, or a combination of tests and analysis.

The Mechanical Engineering Branch of NRC's Office of Nuclear Reactor Regulation prepared an interim technical position on the functional capability of essential piping systems* to serve as a guide for applicants in preparing their functional capability assurance programs. In the interim technical position, the staff indicated that meeting Equation (1) with Level C limits was sufficient assurance of functional capability for components with $D_o/t < 50$. The applicant was to provide additional demonstration for components using Level D limits and for components with $D_o/t > 50$.

During specific plant reviews, applicants submitted other methods of demonstrating functional capability to the NRC. Among these was "Functional Capability Criteria for Essential Mark II Piping" (Ref. 5), which included guidance for components with $D_o/t > 50$.

It is apparent that functional capability assurance requires, in addition to the Code rules, another set of evaluations. An ideal solution to the problem would consist of evidence that meeting the Code rules for piping (with modifications discussed later) would also ensure the functional capability of piping systems.

2.3 Nuclear Regulatory Commission Piping Review Committee Report

Starting in early 1983, the NRC Piping Review Committee reviewed nuclear power plant piping in the context of current regulations, regulatory guides, standard review plans, and other pertinent documents. The results of the review were published in late 1984 and early 1985 in NUREG-1061 (Ref. 6), which consists of five volumes.

---

*Essential piping systems are piping systems that are necessary (1) for safe shutdown of the plant and for maintaining the plant in a safe shutdown condition or (2) for preventing or mitigating the consequence of an accident that could result in potential offsite exposures exceeding the guidelines of 10 CFR Part 100. Piping systems that are not essential do not require a functionality evaluation.

Functional capability of piping is discussed in Section 2.8.5 of Volume 2 of NUREG-1061. By 1984, sufficient earthquake-type-loading test data were available to indicate that earthquake loadings on piping systems, in the absence of high static loads, would not cause "collapse" (large plastic deformations) of piping systems. A staff consultant suggested that functionality capability could be ensured by meeting Equation (1) with Level D limits (lesser of $3Sm$, $2Sy$), provided at least one-half of the stress in Equation (1) came from earthquake-induced loadings.

The Piping Review Committee, at that time, was not ready to endorse the consultant's recommendation and recommended the following:

> The functionality criterion for piping will be maintained. Current ASME Code Class 1 or Class 2 stress evaluation procedures, not to exceed Level C limits, will be used. These limits are similar to those now being used on a case-by-case basis to satisfy the functionality criterion. It is recommended that the upcoming EPRI [Electric Power Research Institute]/NRC pipe tests be evaluated to confirm that position and to determine whether it is appropriate to use the current higher Level D stress limits.

The EPRI/NRC tests have now been completed; see Sections 4 and 5 of this report. The remainder of this report consists of evaluating the EPRI/NRC tests, along with other dynamic loading test data, to determine whether it is appropriate to use the current Level D stress limit for ensuring the functionality of piping systems.

2.4 Relevance of Tests to Piping Functional Capability

A significant aspect of the test data is that, with one exception discussed in Section 4.6, none of the tests resulted in loss of functional capability. Thus, the staff's evaluations are based on the premise that the test data provide lower bounds on combinations of steady-state (e.g., weight) and dynamic loadings that will not cause loss of functional capability. This lower bound premise may introduce conservatisms in the staff's recommendations. But, as will become apparent in the following discussions, the premise leads to a significant relaxation of the present NRC position on functionality.

3 BEANEY DYNAMIC LOADING TESTS ON STRAIGHT PIPE

E. M. Beaney of the Berkeley Nuclear Laboratories in the United Kingdom has conducted a series of dynamic loading tests on straight pipes, on straight pipes with stress concentrations, and on straight pipes with discrete components. The reports by Beaney of particular relevance to functional capability are References 7, 8, 9, and 10.

3.1 Relationship Between Accelerations and Moments

Figure 1 illustrates the test arrangement used by Beaney. Table 1 is a summary of the material types, material yield strengths, and pipe dimensions. A sinusoidal dynamic input was applied to the pipe ends as indicated in Figure 1. The tests of direct interest herein were run with the sinusoidal input frequency equal to the first mode natural frequency of the pipe; that frequency is shown in Table 1. Some tests were run with internal pressure in the pipes, as indicated in Table 1.

Figure 2 shows the test results from Reference 8. The input amplitude was increased to about $3g$; the response acceleration, g_r, at the midspan of the pipe (see Figure 1) was measured. The relationship between moment and response acceleration derived by Beaney is

$$M = [386EI/(4L^2f^2)]g_r \qquad (3)$$

where M = moment at midspan of pipe, in.-lb

E = modulus of elasticity, psi ($3E+7$ psi used herein)

I = moment of inertia of pipe cross section, in.^{4}

L = length of pipe, in. (see Figure 1)

f = frequency of input during testing, Hz

g_r = response acceleration

For example, Equation (3) as applied to Test 1 of Reference 8 gives

$$M = 386 \times 3E7 \times 0.01263/(4 \times 147.2^2 \times 5^2)$$

$$= 67.5 \text{ in.-lb per unit } g_r$$

Having a relationship between M and g_r, the g_r corresponding to the theoretical limit moment, M_L, can be calculated as follows. The limit moment (in.-lb) for straight pipe is

$$M_L = D^2tSy \qquad (4)$$

where D = pipe mean diameter, in.

t = pipe wall thickness, in.

Sy = yield strength of pipe material, psi

and g_{rL}, the response acceleration corresponding to M_L, is

$$g_{rL} = M_L/(M/g_r) \qquad (5)$$

For example, Equations (3), (4), and (5) as applied to Test 1 of Reference 8 give

$$g_{rL} = 0.9642^2 \times 0.03583 \times 43200/67.5 = 21.3$$
$$(g \text{ units})$$

3.2 Comparisons with Theoretical Limit Moments

Figure 2 shows g_{rL} for each of the five tests of Reference 8. It can be seen in this figure that, since g_{rL} corresponds to M_L, the limit moment is an approximate upper bound to the moment that could be sustained in these dynamic loading tests. (Test 5 is anomolous in that the applied moment did not exceed about 65% of the limit moment.)

Equation (1), for zero pressure, $2Sy$ limit, permits the application of a bending moment that is about 1.6 times the limit moment. Thus, the results shown in Figure 2 present a paradox: If applied moments in a piping system are accurately calculated, then Equation (1), with a $2Sy$ limit, does not place any limit on input accelerations.

Of course, to accurately calculate moments due to dynamic loads that are high enough to cause gross plastic response, an elastic-plastic analysis would be required.

3.3 Comparisons with Elastic Analysis

An elastic-plastic analysis of piping systems is within the state-of-the-art. However, in the past and, the staff believes, foreseeable future, for piping system analysis, an elastic analysis has been and will continue to be used and, for earthquake loadings, an elastic response spectrum analysis with $+/-15\%$ peak broadening and not more than 5% damping. Thus, it is pertinent to evaluate Beaney's test results in relation to elastic analysis, as described below.

Beaney's tests were run with an essentially constant frequency input. The input "response spectrum" is a single-value acceleration at the test frequency; peak broadening is meaningless. Because the pipe response is similar to that of a single-degree-of-freedom dynamic structure,

$$g_r = g/2\zeta \tag{6}$$

where g_r = acceleration at pipe midspan

g = input acceleration

ζ = damping factor

Figure 2 shows lines representing 1/2%, 1%, 2%, and 5% damping. It can be seen in this figure that, for low-level input, the responses correspond to about 1% damping. However, for high-level input, the response is much less than that indicated by an elastic analysis, even for 5% damping. It is this aspect of an elastic analysis that makes Code Equation (9) [Equation (1) herein] highly conservative for reversing dynamic loads.

Equation (1), for zero internal pressure, $B_2 = 1.0$ (straight pipe), in conjunction with Equations (3) and (6), can be written as

$$M/Z = S = (M/g_r)g/2Z\zeta \tag{7}$$

For example, Equation (7) as applied to Test 1 of Reference 8 for the highest test level of g ($=3.6$), 2% damping, gives

$$S = 67.5\text{x}3.6/(2\text{x}0.02526\text{x}0.02) = 240{,}500 \text{ psi}, \\ \text{amplitude}$$

For comparison with the Code Level D limit, the $2Sy$ limit (not $3Sm$) will be used because Sy relates directly to limit load theory. The ratio of S to $2Sy$ is thus a direct indication of the test dynamic loadings to the dynamic loadings permitted by the Code with a Level D stress allowable of $2Sy$. For example, for Test 1 of Reference 8, $Sy = 43{,}200$ psi:

$$S/2Sy = 240500/(2\text{x}43200) = 2.78$$

Thus, for Test 1 of Reference 8, the maximum input of $3.6g$ is equivalent to 2.78 times the Code Level D allowable.

Table 2 gives values of $S/2Sy$ for all of References 7, 8, 9, and 10 tests. Because collapse did not occur in any of these tests, the $S/2Sy$ values in Table 2 suggest that, for piping functional capability assurance, Equation (1) with limits of about the following is appropriate, provided the moment used in Equation (1) is almost entirely a reversing dynamic moment.

Analysis Damping, %	Stress Limit
2	$10Sy$
5	$4Sy$

Specifically, the column in Table 2 headed "Sw/Sy" (weight stress/yield strength) supports the use of $10Sy$

(2% damping) or $4Sy$ (5% damping) only if the stress due to weight or other steady-state stress does not exceed about $0.15Sy$.

3.4 Weight Stresses

Table 1 includes a column headed "Test Plane." A "V" in this column indicates that the dynamic loading is in a vertical plane as indicated by Figure 1. With this test arrangement, the weight stress adds to the maximum dynamic moment in the downward-displaced position. An "H" in this column indicates that the actuators were rotated 90° from the plane indicated by Figure 1. With this test arrangement, the maximum weight stress is 90° from the location of maximum dynamic moment.

Figures 3 and 4, which show strain at pipe midspan and deformed shape and permanent strain after tests, respectively, are from Reference 7. As indicated in Table 2, the weight stress at the pipe midspan was $0.11Sy$. This weight stress was sufficient to induce biased strains (Figure 3) and a post-test deformed shape (Figure 4). This magnitude of deformation is well below that which will impair functional capability.

Figure 5, which shows mean strain as a function of input acceleration, is from Reference 8. The column in Table 2 headed "Sw/Sy" indicates the weight stresses at pipe midspan. Other than Test 4, which showed the largest Sw/Sy and the highest mean strain, there is no obvious correlation between Sw/Sy and mean strain. In Test 4 high mean strains of about 1.9% were developed; however, these were not sufficient to indicate any significant loss of functional capability. Figures 3, 4, and 5 serve as a warning that weight and other steady-state stresses must be appropriately limited if Code Equation (9) with limits such as $10Sy$ (2% damping) or $4Sy$ (5% damping) is to be clearly defensible.

In Reference 9, Beaney mentions that "the pipe sags due to the one sided effect of gravity," but does not give quantitative data on the magnitude of the sagging.

Figure 6, which shows deformed shape of upper surface of pipe, is from Reference 10. The buckling indicated in this figure apparently occurred only in Test 16, during which a pipe with $D_o/t = 103/1.5 = 69$ was tested. The pipe was filled with water. The combination of large D_o/t, relatively high weight stress, and relatively low dynamic load input ($1.9g$) led to the incipient buckling as depicted in Figure 6. There is a bit of a mystery that Beaney noted but did not explain: Why did signs of buckling occur in Test 16 but not in Test 15?

The onset of buckling could pose a challenge to maintenance of functional capability. Therefore, the staff recommendations in this report will be "hedged" by limiting the applicability to $D_o/t < 50$, that is, the same D_o/t limit imposed by the Code on the applicability of B-indices.

3.5 Pressure Stresses

Table 2, column headed "Sp/Sy," shows the ratios of circumferential stress due to internal pressure to the pipe material yield strength.

Although internal pressure is significant with respect to pressure boundary evaluation, Beaney's tests suggest that internal pressure has little, if any, significance with respect to functional capability. Pressure stresses are, of course, uniform around the circumference of the pipe and, thus, do not bias the displacement direction, in contrast to weight stresses, which may bias the displacement direction.

Code Equation (9) [Equation (1) herein] includes the pressure term $B_1 P D_o / 2t$; its continued use is expected in any foreseeable Code rule changes for pressure boundary evaluations.

4 ELECTRIC POWER RESEARCH INSTITUTE, NRC, AND GENERAL ELECTRIC COMPANY*TESTS OF PIPING COMPONENTS

4.1 Scope of Tests and Reported Results

A total of 41 component tests were run. The types of components included elbows, tees, reducers, straight pipe, and fabricated branch connections. The tests are described and the test results are given in Reference 11. Reference 11 results were supplemented by data provided in a letter from H. Hwang (General Electric Company) to E. C. Rodabaugh dated October 16, 1991 (available in the author's personal file).

Figure 7 shows a representative test arrangement with an elbow as the test component. Dynamic loadings were applied by motions applied to the sled. Numerous runs were made in each test. The run of main interest to functional capability is (in most tests) an earthquake time history applied to the sled, scaled up to the highest magnitude used in the test.

Reference 11 contains the results of measured moments acting on the components. The measured moments were derived from strain gages placed on the inertia arm (see Figure 7). The inertia arm was sufficiently strong so that it responded elastically in all tests. Thus, comparisons can be made between measured moments and theoretical limit moments.

Of the results given in Reference 11, the most significant with respect to functional capability consists of the calculated stress in the component at the highest magnitude of sled input. These stresses were calculated using an elastic response spectrum analysis. The response spectrum was derived from the time-history input to the sled, using 2% or 5% damping. The analysis is based on +/-15% peak broadening of the so-derived response spectrum and gives the moment acting on the component. The calculated stresses can be compared with the Code Equation (9) stress limit of $2Sy$. If the ratio of calculated stress to $2Sy$ is greater than unity, the test indicates that, for functional capability, an Equation (9) stress limit greater than $2Sy$ is defensible.

4.2 Comparisons with Theoretical Limit Moments

4.2.1 Tests on Straight Pipe

Table 3 is a summary of Reference 11 results for what the staff deems to be essentially straight pipe tests. Only

Tests 33 and 34 were, in fact, tests on straight pipe. However, in the tests of tees and reducers, the plastic response was essentially in the pipes, not in either the tees or the reducers.

The column in Table 3 headed "M_m/M_L" indicates that, as in Beaney's tests on straight pipes, the theoretical limit moment is about as much moment as could be applied when pipes are subjected to very high level, simulated earthquake-type dynamic loads.

For Tests 15 and 34, Reference 11 gives data for several runs. Figures 8 and 9 are plots of Tests 15 and 34. The calculated moment represents a measure of the magnitude of the input, analogous to the g-input of Figure 2. The measured moment represents the response analogous to g_r of Figure 2, convertible by means of Equation (3) to a response moment. As in Figure 2, Figures 8 and 9 indicate a rapid increase in response moment at low-magnitude inputs and a leveling off of response moment at high inputs. Appendix A of Reference 11 includes a column headed "DYN MOM/LIM MOM," where

DYN MOM = maximum measured dynamic moment

LIM MOM = calculated static limit moment

It might seem that DYN MOM/LIM MOM should be the same as M_m/M_L in Table 3. This is approximately so, except for the tee tests. For the tee tests, LIM MOM was calculated in Reference 11 as

$$\text{LIM MOM} = D^2 t Sy/B_{2b} \qquad (8)$$

where B_{2b} is defined by the Code as $0.4(D/2t)^{2/3}$

The largest discrepancy exists for Test 11; DYN MOM/LIM MOM = 2.4 compared with M_m/M_L = 0.65. For Test 11, D = 6.491 in., t = 0.134 in., and Sy = 39.7 ksi. B_{2b} = 3.35 and Equation (8) gives

$$\text{LIM MOM} = 6.491^2 \times 0.134 \times 39.7/3.35 = 67 \text{ in.-kip}$$

The DYN MOM used was that calculated at an imaginary location defined as the "tee center"; DYN MOM = 158 in.-kip. Thus, in Appendix A of Reference 11, DYN MOM/LIM MOM = 158/67 = 2.36. In Test 11, essentially all plastic response was confined to a narrow band of the Schedule (Sch.) 10 branch pipe at its juncture with the tee. Thus, in the staff's view, the ratio of 2.4 shown in the appendix is misleading. From a functional capability standpoint, however, the important aspect is that displacements were not sufficient to cause any loss of functional capability.

---

*Subcontracted by EPRI to evaluate test results.

4.2.2 Tests on Elbows

Table 4 is a summary of Reference 11 elbow tests. The limit moment was calculated as follows:

$$M_L = 0.8h^{0.6}D_o{}^2tSy \text{ for } P = 0 \tag{9}$$

$$M_L = 0.96h^{0.6}D_o{}^2tSy \text{ for } P > 0$$

where h = elbow parameter = tR/r^2

$\quad\quad t$ = elbow wall thickness

$\quad\quad R$ = elbow bend radius

$\quad\quad r$ = elbow cross-section mean radius

The basis for Equation (9), for $P = 0$, is discussed in Reference 3. It is based on an in-plane bending limit moment, $P = 0$, theory developed by Spence and Findlay (Ref. 12). The coefficient of 0.96 for $P > 0$ was suggested by the staff (used in Reference 11) to approximate the increase in moment capacity due to internal pressure.

Insofar as the staff is aware, no closed-form theory exists for an elbow limit moment with $P > 0$, or for an out-of-plane or torsional moment. Existing elastic-plastic, finite-element computer programs might be used; however, to pick up the pressure effect and to distinguish between in-plane closing and in-plane opening, such programs would have to include finite displacement effects. Static, in-plane moment tests show that the moment capacity for in-plane closing is much less than for in-plane opening.

For Tests 3 and 13, Reference 11 gives data for several runs. Figures 10 and 11 are analogous to Figures 8 and 9, which, in turn, are analogous to Figure 2.

Figure 11 shows responses that are similar to those of Beaney's straight pipe tests, that is, rapid rise in response at low-magnitude input, followed by a leveling off of response at high-magnitude input. However, the leveling off occurs at about two times M_L, rather than in the vicinity of M_L. For dynamic equilibrium, the moment capacity of the elbow cannot be exceeded in either the closing direction or the opening direction. Thus, a paradox seems to exist.

For Test 13, the paradox is resolved by considering the actual wall thickness of the elbow that was tested. In a letter from H. Hwang (General Electric Company) to E. C. Rodabaugh dated April 21, 1989 (available in the author's file) regarding dimensional measurements of Reference 11 test components, H. Hwang provided wall thickness measurements of the elbows used in the component tests. The Test 13 elbow was nominally Sch. 40, 0.280-in. nominal wall thickness. The measured thicknesses ranged from 0.327 in. to 0.520 in. with an average wall thickness of 0.425 in. Using the average wall thick-

ness changes the calculated limit moment from 189 to 380 in.-kip. Then, $M_m/M_L = 400/380 = 1.05$.

Thus, Test 13, evaluated using actual average wall thickness rather than nominal wall thickness, indicates that Equation (9) is a good indicator of in-plane dynamic moment capacity for a carbon steel elbow.

The average actual wall thickness of the Test 3 elbow was 0.156 in. compared with the nominal wall thickness of 0.134 in. Using the average wall thickness changes the calculated limit moment from 52.3 to 67 in.-kip. Then, $M_m/M_L = 163/67 = 2.43$. Obviously, the use of actual wall thickness does not explain the seeming paradox for Test 3.

However, evidence that the M_m/M_L ratio for Test 3 is credible can be seen in Figure 12. This figure includes

(1) static in-plane closing moment capacity test data from Reference 3; see Table 5 herein

(2) dynamic in-plane moment capacity test data from Reference 3; see Table 5 herein

(3) dynamic in-plane moment capacity test data from Reference 11; see Table 4 herein

Figure 12 shows that Test 3 results are on the high side of static test data, but are consistent with prior dynamic test data; thus, the Test 3 results are credible.

At the other extreme of M_m/M_L in Table 4, for Test 37, $M_m/M_L = 1.31$. This is consistent with static test data for pipe elbows with zero internal pressure. Test 3 and Test 37 elbows had the same nominal dimensions and were made of the same heat of stainless steel material.

Code Equation (9) [Equation (1) herein], for zero pressure, $2Sy$ limit, permits the application of a bending moment of about $1.6M_L$. Thus, the results for Test 37 indicate that, if the applied moments are accurately calculated, Code Equation (9) with a $2Sy$ limit does not place any limit on dynamic (e.g., earthquake-induced) loads.

4.3 Comparisons with Elastic Analysis

In Section 4.2, measured dynamic moments and limit moments were compared. The staff will now compare calculated stresses with a $2Sy$ stress limit. Calculated moments and/or stresses are given in Appendix B of Reference 11. These calculations are based on elastic response spectrum analyses using either 2% or 5% damping and +/−15% peak broadening. The response spectra used were derived from the time-history inputs to the sled; see Figure 7.

4.3.1 Tests on Straight Pipe

Table 6 is a summary of the results of Reference 11 tests, which the staff deems to be equivalent to straight pipe tests. The staff will use Test 9 as an example to illustrate the significance of Table 6.

For Test 9, 2% damping, the calculated stress amplitude is 589 ksi. The material yield strength is 40.8 ksi. Thus, $S/2Sy = 589/81.6 = 7.22$. Bypassing, until later, the question of weight stress and pressure stress, Code Equation (9) could be written as

$$B_1 PD_o/2t + B_2 M/Z < 14.44 Sy$$

That is, looking only at Test 9, the Code limit of $2Sy$ could be increased to $14.44Sy$, and, since no loss of functional capability occurred in Test 9, the increased stress limit would ensure functional capability.

If the moments were to be calculated using 5% damping, Test 9 indicates the Code Equation (9) limit could be increased to $8.0Sy$, but not necessarily any higher.

A salient point is that the defensible stress limit for Code Equation (9) is highly dependent on how the moments acting on the component are calculated. For Test 9:

Moments Calculated	Defensible Code Equation (9) Limit
Accurately, e.g., by elastic-plastic analysis	$1.3Sy$
By elastic analysis, 2% damping	$14.4Sy$
By elastic analysis, 5% damping	$8.0Sy$

Appendix A of Reference 11 includes a column headed "INPUT X/LEVEL D,"

where INPUT X = calculated stress using linear response spectrum analysis, 2% damping, $+/-15\%$ peak broadening, and actual sled input. Stress = $B_2 M/Z$.

LEVEL D = $3Sm = 60$ ksi.

In the following, for brevity, this ratio is designated as X/D.

In Table 6, the analogous ratio is $S/2Sy$, 2% damping. In the context of a meaningful evaluation of the tests, the staff deems that use of Level D = $3Sm = 60$ ksi is inappropriate. A more meaningful ratio is obtained by using Level D = $2Sy$, where Sy is the yield strength of the material used in the tested component. Because in these tests $2Sy > 60$ ksi, the staff's $S/2Sy$ ratio is always less than X/D.

For example, in Test 34 (pipe test), $X/D = 731/60 = 12.2$, which agrees with the "12" shown in Appendix A of Reference 11. But, for 2% damping, $S/2Sy = 731/(2\times44.5) = 8.2$ as shown in Table 6.

In addition, for those tests that involved tees, INPUT X = $B_2 M/Z$, where, for example in Test 11, B_2 ($=B_{2b}$) = 3.34 was used to calculate the $X/D = 16$ shown in Appendix A of Reference 11. Also, in calculating $X/D = 16$, the calculated moment at the imaginary point at the centerline intersections was used. In its evaluations, since the plastic response was confined to a narrow band of the branch pipe at its intersection with the tee, the staff used $B_2 = 1.0$ for straight pipe with the calculated moment at the branch-pipe-to-tee intersection weld. It thereby obtained $S = 269$ ksi and $S/2Sy = 269/(2\times39.7) = 3.4$ as shown in Table 6.

For all Table 6 tests, X/D and $S/2Sy$ are as follows:

Test No.	9	10	11	12	14	15	16	33	34	40
X/D	21	21	16	27	18	13	30	--	12	22
$S/2Sy$	7.2	7.4	3.4	9.0	6.5	11	20	--	8.2	18

It is apparent that the staff's evaluations of Table 6 tests are significantly more conservative (and, it believes, more realistic) than the X/D ratios in Appendix A of Reference 11. Even so, the staff's evaluations support a significant increase in the present Code Equation (9) limit insofar as functional capability is concerned; for example, the lowest $S/2Sy$ of 3.4 suggests that the Code Equation (9) limit could be increased from $2Sy$ to $6.8Sy$, provided the applied moments are calculated using not more than 2% damping.

4.3.2 Tests on Elbows

Table 7 is a summary of the Reference 11 tests on elbows in the same format as that of Table 6.

The stress was calculated using

$$S = B_2 M/Z \qquad (10)$$

where M was calculated using elastic response spectrum analyses, 2% or 5% damping and $+/-15\%$ peak broadening, and

$$B_2 = 1.3/h^{2/3} \qquad (11)$$

Equation (11) is from NB–3683.7 of the Code; h is the elbow parameter as tabulated and defined in Table 4. The B_2 for each h involved in the tests is as follows:

h	0.41	0.25	0.17	0.11
B_2	2.37	3.27	4.29	5.51

The maximum elastic stress in an elbow (with $h > 1.0$) depends on the moment direction:

Moment Direction	Multiplier of $(1/h)^{2/3}$
In-plane	1.86
Out-of-plane	1.59
Torsion	1.00

The B_2 of $1.3/h^{2/3}$ is intended to represent a conservative estimate of the moment capacity of an elbow subjected to an in-plane closing moment. It is conservative for both out-of-plane and torsion moments. However, an important aspect with respect to the staff's recommendations for functional capability criteria is that they are based on the B-indices as prescribed in the present Code. Any future Code revision that would decrease any of the B-indices might invalidate the staff's recommendations.

The ratios in Appendix A of Reference 11, column headed "INPUT X/LEVEL D," are higher than those in Table 7, $S/2Sy$, 2% damping, because Level D = $3Sm$ = 60 ksi is less than $2Sy$. Although the staff's evaluations are more conservative (and, it believes, more realistic) than those in Reference 11, they still suggest that the present limit on Code Equation (9), for functional capability evaluation, can be increased significantly. The lowest $S/2Sy$ in Table 7, 5% damping, is 5.2. This suggests that the $2Sy$ limit can be increased to $10.4Sy$, even when using 5% damping in calculating the applied moments.

4.3.3 Tests on Other Components

Table 8 is a summary of the Reference 11 tests on other components in the same format as that of Tables 6 and 7.

Reference 11 includes the results of 41 component tests. Tables 6, 7, and 8 contain the results of the staff's evaluations of 33 of these tests. The eight tests not included in Tables 6, 7, and 8, and comments concerning them, are the following:

Test	Comment
21, 22	Tests of lugs on pipe—relevant to pressure boundary integrity but not to functional capability
23	Test of elbow with strut restraint—relevant to support loads but not directly to functional capability
24, 32	Static limit moment tests of elbows—results more or less consistent with Reference 3 static limit moment tests
27	Midfrequency and sinesweep tests of a tee—results for this test not given in Appendix B of Reference 11
28, 29	Water hammer tests—discussed in Section 7 herein

4.3.3.1 American National Standards Institute (ANSI) B16.9 Tees

Tests 36, 38, and 39 in Table 8 are tests of 6x6x6 ANSI B16.9 tees, the same type of components included in Table 6, Tests 9, 10, 11, 12, and 14. However, the Table 8 tests are significantly different from the Table 6 tests, as illustrated by the following sketch.

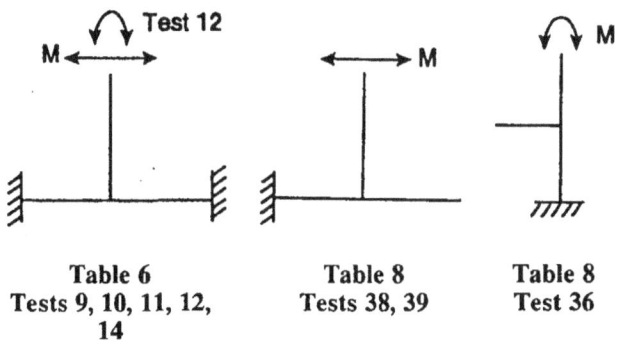

Table 6	Table 8	Table 8
Tests 9, 10, 11, 12, 14	Tests 38, 39	Test 36

In Tests 36, 38, and 39, plasticity and eventual fatigue failure occurred in the body of the tees. Thus, it is deemed appropriate to evaluate these tests using the B_{2b} (Tests 38 and 39) or B_{2r} (Test 36) specified in the Code:

$$B_{2b} = 0.4(R/T)^{2/3} = 2.02; \quad B_{2r} = 0.5(R/T)^{2/3} = 2.52$$

where R = mean radius of attached pipe (3.1725 in.)
T = nominal wall thickness of attached pipe (0.280 in.)

4.3.3.2 Tests 18 and 20, Fabricated Branch Connections

The staff's evaluations of Tests 18 and 20 require a more detailed explanation. The Code-specified B_{2b} index for branch connections per NB–3643 (see NB–3683.8) is

$$B_{2b} = 1.5 \times 3(R/T)^{2/3}(r/R)^{1/2}(t/T)(r/r_p) \qquad (12)$$

where R = mean radius of run pipe

T = nominal wall thickness of run pipe

r = mean radius of branch pipe

t = nominal wall thickness of branch pipe

$r_p \cong$ radius to outside of nozzle

For use in Code Equation (9),

$$S = B_{2b}(M_b/Z_b) \qquad (13)$$

where M_b = moment applied to branch

Z_b = section modulus of branch pipe

Test 18

Test 18 of a pad-reinforced fabricated tee poses a problem because B-indices for pad-reinforced branch connections are not given in the Code. However, the staff believes that B_{2b} for Test 18 can be bounded by using Equation (12) with $T = 0.322$ in. (Sch. 40 run pipe) as an upper bound and $T = 0.322$ + pad thickness = 0.644 in. as a lower bound. For $T = 0.322$ in., Equation (12) gives

$$B_{2b} = 1.5(4.1515/0.322)^{2/3}(2.1315/4.1515)^{1/2}$$
$$(0.237/0.322)(2.1315/2.25) = 4.12$$

Changing only the T of 0.322 to 0.644 in. gives $B_{2b} = 1.30$. In its evaluation of Test 18, the staff used an average B_{2b} of 2.7.

For Test 18, Run 6, Appendix B of Reference 11 gives

M = 915 in.-kip, 2% damping

M = 542 in.-kip, 5% damping

These moments are used for M_b in Equation (13) to give

S = 2.7x915/3.21 = 770 ksi for 2% damping

S = 2.7x542/3.21 = 456 ksi for 5% damping

For Test 18, $B_{2b} = 4.12$ was used in Reference 11 without an explanation of its basis (perhaps coincidentally, $B_{2b} = 4.12$ can be obtained from Equation (12) for an unreinforced fabricated tee). In Reference 11, the Code-prescribed $Z_b = 3.21$ in.^{3} was used.

Test 20 (see Figure 13)

As applied to Test 20, Equation (12) gives

$$B_{2b} = 1.5(6.1875/0.375)^{2/3}(2.1315/6.1875)^{1/2}$$
$$(0.237/0.375)(2.1315/2.25) = 3.416$$

For Test 20, Run 7, Appendix B of Reference 11 gives

M = 724 in.-kip, 2% damping

M = 410 in.-kip, 5% damping

These moments are used as M_b in Equation (13) to give

S = 3.416x724/3.21 = 770 ksi for 2% damping

S = 3.416x410/3.21 = 436 ksi for 5% damping

For Test 20, $B_{2b} = 7.79$ and $Z_b = 5.9$ in.^{3} were used in Reference 11 without an explanation of their basis. They are obviously not in accordance with the Code.

X/D in Appendix A of Reference 11 and $S/2Sy$, 2% damping, in Table 8 are as follows:

Test No.	18	20	36	38	39
X/D	20	16	15	20	21
$S/2Sy$	7.2	7.9	9.9	15	16

As its evaluations in Tables 6 and 7, it is apparent that the staff's evaluations in Table 8 are more conservative than those in Reference 11. Even so, the staff's evaluations suggest that, for functional capability evaluations, the present limit on Code Equation (9) can be increased significantly. The lowest $S/2Sy$ in Table 8, 5% damping, is 4.3. This suggests that the $2Sy$ limit might be increased to $8.6Sy$, even when 5% damping is used in calculating the applied moments.

4.4 Weight Stresses

Tables 6, 7, and 8, column headed "Sw/Sy," show weight stresses as ratios to yield strength, Sy. These ratios, except for Tests 30 and 37 (elbows), are not more than 0.08. Thus, they are of limited usefulness with respect to establishing a reasonable bound on weight stresses combined with reversing dynamic stresses.

Tests 30 and 37 are discussed in Section 4.6.

4.5 Pressure Stresses

Tables 6, 7, and 8, column headed "Sp/Sy," show the nominal pressure stresses, $PD_o/2t$, as ratios to yield strength, Sy.

Although internal pressure is significant with respect to pressure boundary evaluation, the data do not suggest any decrease in functional capability for Sp/Sy ratios up to 0.48. Indeed, as discussed in Section 4.6, internal pressure in elbows appears to increase their moment capacity. This "inverse" pressure effect is also apparent in static tests on elbows; see Reference 3.

4.6 Tests 30 and 37

Tests 30 and 37 were in-plane moment tests on nominally identical elbows. The test arrangement is shown in Figure

14. The assemblies were "tuned" (height of vertical arm, magnitude and location of weights, etc.) so that the first mode response frequency was about 1.4 Hz. The time-history input was adjusted so that the peak of the input response spectrum was at about 1.3 Hz. The adjustment was made by expanding the time of the time-history input; a run for Tests 30 and 37 lasted about 110 sec, rather than the about 20 sec for other simulated earthquake inputs.

The elbows were from the same heat of stainless steel material with Sy = 34 ksi. They were 6 nominal pipe size (NPS), Sch. 10, 9-in. bend radius.

The weight stress, at the mid-arc of the elbows, was 10.74 ksi for both elbows with Sw/Sy = 0.32.

The only apparent testing differences were the following:

Test/Run	Pressure	S, 2% Damping
30/4	400 psi	620 ksi
37/5	0	651 ksi

Test 30 was ended when a fatigue failure occurred. Some permanent displacements occurred (not quantified in Reference 11) during the test runs, but it is believed that none of these deformations were sufficient to reduce functional capability.

Test 37 consisted of low-level Runs 1 and 2 and then

Run	3	4	5
S, 2% damping, ksi	79	324	651

During Runs 1 through 4, displacements were relatively small. During Run 5, after about 45 sec into the run, the assembly began to ratchet-displace in the elbow closing direction. The test was terminated at about 72 sec into Run 5 because the displacements were becoming large and increasing rapidly with time.

At the termination of Test 37, the upper end of the inertia arm had displaced several feet and, if the test had been continued for a few more seconds, the displacements would probably have increased to the extent that the elbow cross section would have significantly decreased; that is, functional capability would have been lost.

Although about 5% higher loadings were used in Test 37, Run 5, than in Test 30, Run 4, the staff believes the major difference is that Test 37 was run at zero pressure, while Test 30 was run at 400-psi pressure. The measured moments in Test 37, Run 5, and Test 30, Run 4, were 57 and 112 in.-kip, respectively. This rather directly indicates the increase in moment capacity due to an internal pressure of 400 psi with Sp/Sy = 0.24.

Test 37 is a direct indication that a weight load (10.74-ksi weight stress) that would not cause collapse by itself, in combination with high reversing dynamic loads (S, 2% damping = 651 ksi), does cause collapse.

However, Test 37 must be looked at in light of the following:

(1) The test was meant to be an extreme evaluation of the concept that reversing dynamic loads do not cause collapse. The pipe parameters selected for this extreme case included in-plane moments (weakest direction), zero pressure (worst case for pressure), and thin-walled Sch. 10 pipe, for which D_o/t = 49 (high D_o/t and pronounced elbow effects with h = 0.11). The weight stress was such that Sw/Sy was 0.32 (significantly higher than the usual weight stresses in piping systems). The very low test frequency of about 1.3 Hz may have contributed to the collapse in that 1.3 Hz gives the assembly more time to displace before reversal of dynamic load occurs.

(2) The dynamic loads were very high; that is, $S/2Sy$, 2% damping, was 9.6, which is 9.6 times the present Code Level D limit.

(3) Test 37 was a component test. In a piping system, additional plastic hinges would have to develop before large plastic displacements could occur.

Nevertheless, Test 37 constitutes a "red flag" to indicate that appropriate control must be placed on steady-state loadings to avoid the possibility of loss of functional capability during application of high reversing dynamic loads.

5 ELECTRIC POWER RESEARCH INSTITUTE, NRC, AND GENERAL ELECTRIC COMPANY TESTS OF PIPING SYSTEMS

Two piping systems, identified as System 1 and System 2, were tested. The system configurations, testing, and results are given in Reference 13. Reference 13 results are supplemented by additional data provided in a letter from H. Hwang (General Electric Company) to E. C. Rodabaugh dated November 14, 1991 (available in the author's file).

5.1 Piping System Configurations and Materials

Figures 15 and 16 show the configurations of Systems 1 and 2, respectively.

System 1 was made of carbon steel (A106–B). It is characterized in Reference 13 as follows: "[System 1] was relatively balanced with regard to dynamic strain such that several different locations had cyclic plastic strains of about the same magnitude."

System 2 was made of stainless steel (Type 316). It is characterized in Reference 13 as follows: "[System 2] had unbalanced stresses with a single high-stress location where failure was predicted to occur while the remainder of the piping system was at a relatively lower stress."

5.2 Loadings

Both Systems 1 and 2 were tested with an internal pressure of 1000 psi.

Reference 13 describes the various time-history inputs used in the system tests. From the standpoint of functional capability, the highest input is of primary significance for both systems; the highest input was associated with "Time History B," with all sleds acting in unison.

5.3 Comparisons with Theoretical Limit Moments

In principle, dynamic moments at any location in a piping system cannot exceed the moment capacity at that location. This aspect of dynamic loading tests is discussed in Sections 3.2 and 4.2. Comparisons of test measured moments with calculated moments are shown in Figures 1 and 8 through 11 and Tables 3, 4, and 5.

Reference 13 gives measured and calculated moments at three locations as shown below:

Ref. 13 Table	System	Location
5–5	1	Node 72, 6 NPS short-radius elbow
6–9	2	Node 6, 12x4 NPS nozzle, see Figure 13
6–11	2	Node 52, 6 NPS, Sch. 40 pipe

Table 5–5		Table 6–9		Table 6–11	
Mm	Mc	Mm	Mc	Mm	Mc
66.1	92.5	39	53	110	150
330	634.7	78	96	255	293
725	3228.4	119	334	576	915
717	4994.1	156	572	681	1551
		235	1091	837	2658

where Mm = measured moment, in.-kip

Mc = calculated moment, in.-kip, using 2% damping, +/–15% peak broadening

The data were sufficient so that plots of Mm versus Mc, analogous to Figures 8 through 11, could be made. However, such plots are of little value because of the major uncertainties discussed below.

Reference 13 does not describe how the moments were measured. However, in a letter from H. Hwang (General Electric Company) to E. C. Rodabaugh dated November 14, 1991 (available in the author's file), Hwang stated that the measurement devices shown in Figure 17 were used in both Systems 1 and 2. To the extent that applied moments do not exceed the yield moment of the 6 NPS, Sch. 160 pipe on which strain gages were mounted, the strain measurements can be used to calculate measured moments (e.g., at Node 72 in System 1).

The yield strength of the Sch. 160 pipe used in System 2, according to Table 6–1 of Reference 13, is 31.3 ksi. The yield moment of the load measurement device (Figure 17) is then

$$My = SyZ = 31.3 \times 17.81 = 557 \text{ in.-kip}$$

Thus, the values of Mm from Table 6–11 greater than 557 in.-kip may reflect yielding of the measurement device and not be an accurate indication of the measured moment. It is this uncertainty that makes comparison with the limit moment of straight pipe questionable.

The moments cited from Reference 13, according to the letter from H. Hwang to E. C. Rodabaugh dated November 14, 1991, are resultant moments; that is,

$$M = (Mx^2 + My^2 + Mz^2)^{1/2}$$

The moment capacity of elbows (Table 5–5) and nozzles (Table 6–9) are significantly dependent on the orientation of the applied moments. It is this uncertainty that makes comparisons with limit moments of elbows or nozzles questionable.

Accordingly, no meaningful comparisons can be made between Reference 13 measured moments and either limit load theory or tests or the component tests of Reference 11 discussed herein in Section 4.

5.4 Comparisons with Elastic Analysis

Reference 13 gives the following calculated moments:

Ref. 13 Table	System	Location	Sy, ksi	Mc, in.-kip, for Damping of 2%	5%
5–5	1	Node 72, 6 NPS short-radius elbow	43.8	4994	----
6–9	2	Node 6, 12x4 NPS nozzle	35.7	1091	775
6–11	2	Node 52, 6 NPS Sch. 40 pipe	35.0	2658	1860

where Mc is the moment calculated by an elastic response spectrum analysis, 15% peak broadening, damping of 2% or 5% as indicated. These values are taken from the indicated Reference 13 tables under the columns "Full Sled-4 ARS" for System 1 and "FULL UNIF" for System 2. Sy is the material yield strength according to Reference 13, Tables 5–1 and 6–1. For the 12x4 nozzle, Sy is for the 12 NPS pipe.

The staff evaluated these calculated moments in a manner analogous to that for Tables 2, 6, 7, and 8, that is, develop ratios of $S/2Sy$, where $S = B_2M/Z$ and Sy = yield strength of the material. The results are summarized as follows:

Ref. 13 Table	Sy, ksi	B_2	Z, in.³	2% Damping S, ksi	$S/2Sy$	5% Damping S, ksi	$S/2Sy$
5–5	43.8	4.29	8.50	2520	29	---	(16)
6–9	35.7	3.42	3.21	1162	16	826	12
6–11	35.0	1.00	8.50	313	4.5	219	3.1

For Table 5–5, a significant uncertainty existed concerning the actual wall thickness of the short-radius elbow. This uncertainty goes back to the measured wall thicknesses that were available for Component Test 13, which indicated that the average actual wall thickness was 0.425 in. rather than the nominal wall thickness of 0.280 in. However, in a letter from W. P. Chen (Energy Technology Engineering Center) to E. C. Rodabaugh dated December 19, 1991 (available in the author's file), Chen provided the measured wall thicknesses of elbows in Systems 1 and 2. For the short-radius elbow at Node 72, the average wall thickness is 0.310 in., which is only 11% more than the nominal wall thickness.

Table 5–5 of Reference 13 does not give a calculated moment for 5% damping. However, Table 2–1 in the Executive Summary of Reference 13 gives INPUT X/LEVEL D = 24.0 for 5% damping. The $S/2Sy$, shown in parentheses, was obained using $S/2Sy$ = 24x60/(2x43.8) = 16.4.

Table 6–9 of Reference 13 pertains to the evaluation of a nozzle that, according to H. Hwang (General Electric Company) in a letter to E. C. Rodabaugh dated November 14, 1991 (available in the author's file), was dimensionally the same as that used for Component Test 20; see Figure 13 herein. The calculation of B_2 = 3.42 is discussed herein in Section 4.3.3.2.

Table 6–11 of Reference 13 pertains to the evaluation of straight pipe, with no complications. Staff ratios of $S/2Sy$ and Reference 13 ratios of $S/3Sm$ are as follows:

Ref. 13 Table	$S/2Sy$ 2%	5%	$S/3Sm$ 2%	5%
5–5	29	16	42	24
6–9	16	12	21	15
6–11	4.5	3.1	5.2	3.7

As its evaluations in Tables 6, 7, and 8, it is apparent that the staff's evaluations are more conservative than those in Reference 13, largely because the staff incorporated the material property, Sy, rather than using Sm = 20 ksi. Even so, the staff's evaluations indicate that, for functional capability, the present Code Equation (9) limit of $2Sy$ could be increased to $9Sy$ for 2% damping or to $6Sy$ for 5% damping.

5.5 Weight Stresses

Figures 18 and 19 show weight stresses, S_{WT}, for Systems 1 and 2. Reference 13 gives no further information on weight stresses. However, in a letter from H. Hwang (General Electric Company) to E. C. Rodabaugh dated November 14, 1991 (available in the author's file), Hwang provided the following information:

(1) $S_{WT} = B_2 Mw/Z$, or $B_{2b} Mwb/Z$, or $B_{2r} Mwr/Z$

where B_2, B_{2b}, B_{2r} are Code indices

Mw = resultant moment due to weight at elbows

Mwb = resultant moment due to weight on branch of tees and at Node 8, System 1, and Node 6, System 2

Mwr = moment due to weight on runs of tees

Z = pipe section modulus

(2) Mw, Mwb, and Mwr were calculated by analyses of the piping systems with weight loading, including the weight of water in the systems.

(3) S_{WT} is in units of ksi.

For the highest weight stress in each system:

System	Fig.	Node	Component	B_{2b}	S_{WT}, ksi	Sy, ksi	S_{WT}/Sy
1	18	8	Vesselet	3.48	8.5	45.4	0.19
2	19	6	Nozzle	3.61	10	35.7	0.28

Thus, the staff concludes from the two piping system tests that weight stresses of up to about 0.25Sy in combination with the high reversing dynamic loads such as those applied in the tests will not impair the functional capability of piping systems.

5.6 Pressure Stresses

Systems 1 and 2 were tested with an internal pressure of 1000 psi. The pressure stresses, $PD_o/2t$, of particular relevance are the following:

System/ Node	Sy, ksi	Dimensions	Sp, ksi	Sp/Sy
1/72	43.8	6 NPS, 0.280-in. wall	11.3	0.26
2/52	35.0	6 NPS, 0.280-in. wall	11.3	0.32
2/6	35.7	4 NPS, 0.237-in. wall	8.99	0.25

From the standpoint of pressure boundary integrity, using an internal pressure of 1000 psi was appropriate, although using a higher internal pressure would have been even more appropriate. However, from the standpoint of functional capability, using zero internal pressure might have been more bounding; that is, the elbows in the systems would have had lesser moment capacity.

6 OTHER PIPING SYSTEM TESTS

Piping systems, other than the two systems discusssed in Section 5, were tested using dynamic loadings. Table 9 (Refs. 14–20) identifies the tests the staff reviewed for this report.

The major purpose of reviewing other piping system tests was to see if any of those tests might invalidate conclusions drawn from the evaluations in Sections 3, 4, and 5. The Hanford Engineering Development Laboratory (HEDL) tests (Ref. 14) are the most significant because large displacements occurred to the extent that functional capability was threatened. The HEDL tests are described in Section 6.1. Except for the HEDL tests, the other piping system tests, like the two system tests discussed in Section 5, did not result in any threat to functional capability and, like most tests in Sections 3, 4, and 5, provide lower bounds on combinations of steady-state (weight) and dynamic loadings that will not cause loss of functional capability.

For direct comparison with the tests discussed in Sections 3, 4, and 5, it would be ideal to have elastic response spectrum analyses using 2% or 5% damping and +/– 15% peak broadening. If the yield strength of the material at the highest stress location were available, calculation of $S/2S_y$ in direct analogy to those calculated in Sections 3, 4, and 5 would be possible. Of the cited references (Refs. 14–20), none provide exactly what would be needed. However, each of the references does provide elastic analyses, which are sufficient for the major purpose of determining if any of these tests invalidate conclusions drawn from the evaluations in Sections 3, 4, and 5.

6.1 Hanford Engineering Development Laboratory Tests (Reference 14)

The HEDL piping system has undergone numerous tests, starting in 1979 with design verification tests and continuing to 1985 (see Table 1–1 of Reference 14). The tests of primary interest in this report are identified in Reference 14 as "modified four-support configuration" tests and are discussed below.

The configuration is shown in Figure 20a. The system was subjected to sinusoidal input at a frequency of 2 Hz, with step increases of maximum acceleration levels up to 2.8g. The 2–Hz input frequency coincides with the measured first mode natural frequency of 2 Hz. The calculated first mode frequency is 2.14 Hz. The system was not pressurized.

Figure 20b shows the displacement of point A as a function of input acceleration. At the end of the 2.8g test, point A was permanently displaced 18 in. in the positive Z direction. Plastic hinges developed at points B and C, but there was no significant reduction in cross-sectional flow area.

Directional changes in the piping system were made by cold-bending the pipe to a bend radius of 3 in. The B_2 index for the bends is $1.3h^{2/3} = 1.3/1.142^{2/3} = 1.19$. The ratio of the in-plane, closing, limit moment, using Equation (9), to the straight pipe limit moment with the same S_y, using Equation (2), is 1.17. Thus, even ignoring the cold-bending effect on yield strength, the bends had only a little less moment capacity than the straight pipe. Cold-bending significantly increases the yield strength of an austenitic stainless steel material. Thus, the bends would be expected to be stronger than the straight pipe and, indeed, the hinge at point B formed in the straight pipe. The significance of this lies in generalization of the "no significant loss in flow area." If, for example, ANSI B16.9 elbows with a bend radius of 1.5 in. had been used in the HEDL system, and if the elbow material yield strength were not greater than that of the pipe, the hinge at point B might form in the elbow with incipient loss of flow area.

Reference 14 gives the results of an elastic response spectrum analysis using 10% damping, no peak broadening. The response spectrum used is shown in Figure 5–1 of Reference 14, that is, a response spectrum for a sinusoidal input at 2 Hz. The highest stress in the system was calculated to occur at the bend at point B in Figure 20b. At an input of 0.36g, the highest stress was calculated to be 60 ksi. The yield strength of the pipe material is stated to be 40 ksi. For a maximum input of 2.8g,

$$S/2S_y = 60 \times 2.8/(0.36 \times 2 \times 40) = 5.83 \quad 10\% \text{ damping}$$

An estimate of $S/2S_y$ for 5% damping is needed. There are indications in Reference 14 that the system essentially responded as a single-degree-of-freedom system to the 2–Hz sinusoidal input. Then, the calculated stress for 5% damping would be two times that for 10% damping, leading to

$$S/2S_y = 5.83 \times 2 = 11.7 \quad 5\% \text{ damping}$$

This $S/2S_y$, rounded off to 12, is shown in Table 9. Using the hypothesis of single-degree-of-freedom response, peak broadening is meaningless and the $S/2S_y$ ratios for the first three entries in Table 9 are deemed to be reasonably comparable.

Reference 14, Figure 5–2, indicates weight stresses did not exceed about 10 ksi, $S_w/S_y < 0.25$. Thus, the HEDL tests cannot be used to defend high reversing dynamic loads in combination with steady-state stresses higher than $0.25S_y$. However, the HEDL tests obviously support a $4S_y$ limit on Code Equation (9). Indeed, from only these tests, a limit of about $20S_y$ is defensible.

6.2 References 15–20 Tests

The piping system tests in References 15 through 20 did not result in any threat to functional capability. In the following sections, the staff briefly describes how the $S/2Sy$ ratios shown in Table 9 were derived from the cited references and what information on weight stresses can be gleaned from the references.

6.2.1 Reference 15

Figure 4 of Reference 15 is a response spectrum for 2% damping. It indicates that the maximum test input was about four times that required to produce a maximum calculated stress of $2.4Sh$. For the A106–B material, $Sh = 15$ ksi, $2.4Sh = 36$ ksi. The yield strength, Sy, of the pipe material is not given. Using a typical Sy of 45 ksi leads to

$$S/2Sy = 4\times36/(2\times45) = 1.6 \quad 2\% \text{ damping}$$

The effect of peak broadening is not discussed in Reference 15. Weight stresses are not given, but Sw/Sy was probably less than 0.1.

6.2.2 Reference 16

Reference 16 states that the piping system without branches withstood seismic inputs that were approximately four times the input required to produce a calculated stress equal to the Level D stress limit for Class 2 piping. It appears that the calculated stresses are from an elastic response spectrum analysis using 3% damping and, probably, no peak broadening. Although Reference 16 does not give important details, it appears that this would translate to an $S/2Sy$ between 2 and 4; a value of 3 is shown in Table 9. This $S/2Sy$ is very approximate.

Reference 16 does not give analogous information for the piping system with branches. Weight stresses are not given, but Sw/Sy was probably less than 0.1.

It appears that for both systems, Reference 16 tests were low-level tests relative to References 13, 14, and 17 tests.

6.2.3 Reference 17

Reference 17 gives the results of elastic response spectrum analyses using 5% damping, no peak broadening. Table 2–1 of Reference 17 shows the following:

$$S/3Sm = 30/1.4 = 21 \text{ for 3 NPS system}$$
$$S/3Sm = 30/2.0 = 15 \text{ for 6 NPS system}$$

where $3Sm = 60$ ksi

The yield strength, Sy, for the A106–B pipe material is not given for the 3 NPS system. Using a typical Sy of 45 ksi leads to

$$S/2Sy = 21\times60/(2\times45) = 14 \quad 5\% \text{ damping, 3 NPS system}$$

The yield strength of the A106–B material for the 6 NPS system, according to Table 6–2 of Reference 17, is 54 ksi. Thus,

$$S/2Sy = 15\times60/(2\times54) = 8.3 \quad 5\% \text{ damping, 6 NPS system}$$

Reference 17 indicates the weight stresses are not more than 10 ksi; Sw/Sy was less than about 0.2.

6.2.4 Reference 18

Reference 18 gives the results of an elastic response spectrum analysis using 2% damping, +/–15% peak broadening. The maximum calculated $S/3Sm$ was 2.5; $3Sm = 60$ ksi. The yield strength, Sy, of the austenitic stainless steel piping material is not given. Using a typical Sy of 35 ksi leads to

$$S/2Sy = 2.5\times60/(2\times35) = 2.1 \quad 2\% \text{ damping}$$

Weight stresses are not given, but Sw/Sy was probably less than 0.2.

6.2.5 Reference 19

Reference 19 gives the results of elastic response spectrum analyses using (probably) 3% damping, no peak broadening. Calculated stresses are summarized in Table 7 of Reference 19. The highest calculated stress is 25.5 ksi at location "QA100" for Test T41.21.2. This test is for the "KWU support configuration" at an input excitation of "300%SSE." The maximum input was 800%SSE; thus, $S(\text{nom}) = 25.5\times8/3 = 68$ ksi. Stresses in Reference 19 were calculated using M/Z, not B_2M/Z. Location "QA100" is at an 8 NPS, 0.535–in. wall, 12–in. bend radius elbow for which $B_2 = 2.42$. Thus, $S = B_2M/Z = 2.42\times68 = 165$ ksi at 800%SSE input. The yield strength, Sy, of the austenitic stainless steel pipe material is not given. Using a typical Sy of 35 ksi leads to

$$S/2Sy = 165/(2\times35) = 2.4 \quad 3\% \text{ damping}$$

Weight stresses are not given, but Sw/Sy was probably less than 0.2.

6.2.6 Reference 20

Reference 20 gives the results of a linear elastic time history analysis using (probably) damping equivalent to about 1% in a response spectrum analysis.

The calculated results are summarized in Reference 20 as follows:

Maximum test input	1895 gal (1.93g)
Maximum allowable input for Code Equation (9) = 60 ksi	240 gal (0.24g)

Figures 6 through 10 of Reference 24 indicate that the yield strength of the pipe material was about 25 kg/mm = 36 ksi. Thus,

$$S/2Sy = (1895/240)60/(2 \times 36) = 6.6$$

The $S/2Sy$ of 6.6 in Table 9 is shown in parentheses because a response spectrum analysis was not available. It may be that, on the basis of a response spectrum analysis, Reference 20 tests were low-level tests relative to References 13, 14, and 17 tests.

Reference 20 cites a weight stress of 0.1 kg/mm = 0.14 ksi, presumably at the location at which S is maximum. No other weight stresses are given, but Sw/Sy probably did not exceed 0.2 at any location.

6.3 Summary of Other Piping System Tests

Results of the other piping system tests do not invalidate the conclusions drawn from the evaluations in Sections 3, 4, and 5 of this report. In particular, the HEDL tests (Ref. 14) support a limit of $4Sy$ on Code Equation (9). None of the other piping system tests provide a defense of steady-state stresses greater than about $0.25Sy$ when combined with reversing dynamic stresses of $4Sy$.

7 OTHER DYNAMIC LOADS

Dynamic loads applied in the tests discussed in Sections 3 through 6 were rapidly reversing in nature. The "rapidly" is quantified as dominant reponses of 2 Hz or more. Provided the dominant response is not less than about 2 Hz, these tests support, in regard to asssurance of functional capability, an increase in the Code Equation (9) Level D limit from $2S_y$ to $4S_y$, with steady-state stresses up to about $0.25S_y$.

Other dynamic loads are the result of the following:

- fluid hammer

  - fluid pressure waves

  - slug flow

- relief-valve actuation

  - steady-state forces

  - short-time effects

- postulated pressure boundary breaks

- vibrations (e.g., piping connected to a reciprocating pump)

The question discussed in the following sections is: Can the Code Equation (9) Level D limit be increased when other dynamic loads are applied to piping systems, either alone or in combination with rapidly reversing dynamic loads?

7.1 Fluid Hammer

A part of the EPRI, NRC, and General Electric Company program consisted of water-hammer tests. These tests are described in Section 7, "Pipe System Water Hammer Tests," of Reference 13.

Five piping systems were tested; the tests were identified as Test 28, Test 29, MS–1, MS–2 Runs 1–5, and MX–2 Runs 6 and 7. Water-hammer tests consisted of

(1) piping systems filled with water, sudden pressure increase at one end of system: "solid water-hammer load"

(2) piping systems partially filled with water, sudden pressure increase at one end of system: "slug-type loading"

The conclusions quoted from Reference 13 are the following:

> 7.3.4 Water Hammer Test Conclusions
>
> In general solid water wave load, because of quick load reversal, does not cause pipe collapse, even when the calculated moment exceeds the limit moment.
>
> Strut failure due to water hammer can occur, but in the test the failure load exceeded 10 times of its rated load.
>
> Slug type loading of long duration (simulating static loads) can cause pipe "collapse."

A fluid (e.g., steam or water) pressure wave load could be caused by closing of a valve or slamming of a check valve. Time-history analyses are used to evaluate such loads, and damping is not very significant. The staff agrees with the conclusion in Reference 13 that pressure wave loads are appropriately included with other rapidly reversing dynamic loads.

As indicated by the third conclusion in Reference 13, slug flow may produce collapse and thus constitutes a threat to functional capability. No increase in Code Equation (9) can be defended.

Slug flow is, of course, difficult to anticipate in the design stage. Designs should include drains and vents, and operating procedures should be implemented so that the possibility of slug flow is minimized.

7.2 Relief-Valve Actuation

The steady-state thrust (e.g., acting for one or more seconds) should be evaluated as equivalent to a weight stress.

The time-variable effects would depend on whether there is any slug flow. However, whether there is slug flow or not, the information is insufficient to defend any increase in Code Equation (9) Level D limits for those portions of piping systems on which the relief valve is mounted.

In boiling-water reactors, relief-valve actuation may cause building vibration. The effect of this building-filtered vibration on piping systems is appropriately included with other rapidly reversing dynamic loads.

7.3 Postulated Pressure Boundary Breaks

A concern is whether the postulated pressure boundary break might cause loss of functional capability of piping systems other than the system in which the break is postulated.

Because the break effects will be "filtered" at other piping systems, the staff believes that the effects of postulated breaks can be considered to be rapidly reversing for the purpose of evaluating piping systems other than the system in which a break is postulated.

7.4 Vibrations

Vibrations, such as those induced by attached equipment or fluid flow, are difficult to anticipate in the design stage. The staff believes that such vibrations are best evaluated during preoperational testing.

8 SUMMARY AND LIMITATIONS

The objective of this report is to examine present Code rules and potential changes in Code rules to see if they are sufficient to ensure maintenance of functional capability.

Stresses calculated by using Code Equation (9) [Equation (1) herein] are limited as indicated in Section 2.1 of this report.

As indicated in Section 2.2 of this report, the staff believes that for static loadings, meeting Code Equation (9) with Level D limits does not adequately demonstrate functional capability. However, as discussed in the previous sections of this report, the results of many dynamic tests show that the functionality of piping systems has been maintained at equivalent stress levels significantly higher than Level D limits. The following sections summarize the findings and the limitations for ensuring piping functionality.

8.1 Reversing Dynamic Loads

Reversing dynamic loads are those due to earthquakes and building-filtered loads such as those due to vibration of buildings caused by relief-valve actuation in boiling-water reactors.

The test data evaluated in Sections 3, 4, 5, and 6 of this report are relevant to this type of dynamic loading.

A significant aspect of the test data is that, with one exception discussed in Section 4.6, none of the tests resulted in loss of functional capability. Thus, the staff's evaluations are based on the premise that the test data provide lower bounds on combinations of steady-state (e.g., weight) and dynamic loadings that will not cause loss of functional capability. This lower bound premise may introduce conservatisms in the staff's recommendations. But, as will become apparent in the following discussions, this premise leads to a significant relaxation of the present NRC position on functionality; that is, present Code Level D limits ensure piping functionality provided steady-state stresses do not exceed $0.25Sy$ and the dynamic loadings are similar to those induced by earthquake internal loadings.

8.1.1 Method of Calculating Mi in Code Equation (9)

The moment, Mi, represents both steady-state (e.g., weight) loads and dynamic loads. Values of Mi are obtained by analyses of piping systems. In the past, the dynamic portion of Mi has been obtained by an elastic response spectrum analysis with $+/-15\%$ peak broadening and as low as 0.5% damping. The present trend is to

use 2% and up to 5% damping. Thus, the staff's evaluations are focused on 2% or 5% damping.

However, it is within the state-of-the-art to more accurately calculate the dynamic portion of Mi using an elastic-plastic analysis. The approach used by the staff in Sections 3.2, 4.2, and 5.3, "Comparisons with Theoretical Limit Moments," was to look at Code Equation (9) with the thought that Mi might be more accurately calculated. The staff concludes that, if Mi is accurately calculated, Code Equation (9), with a $2Sy$ limit, is not conservative.

If elastic-plastic analyses of piping systems in nuclear power plants become routine, the staff believes that, for ensuring piping functionality, a revised set of guidelines might be needed for NRC's acceptance of such analyses. Thus, its recommendations discussed herein apply only to elastic response spectrum analyses.

8.1.2 Summary of $S/2Sy$ Evaluations

Table	$S/2Sy$ 2% Damping Min.	Max.	Avg.	5% Damping Min.	Max.	Avg.	Sw/Sy Max.
2	0.62	5.5	3.5	0.25	2.2	1.4	0.15
6	3.4	20	10	2.2	10	5.6	0.06
7	9.1	24	13	5.2	15	7.4	0.32
8	7.2	16	11	4.3	8.4	6.4	0.08
N/A*	4.5	29	16	3.1	12	7.6	0.28

In the staff's judgment, the averages of $S/2Sy$ are reasonable indicators of lower bounds on functional capability, since functional capability was not lost in any tests other than Test 37 of Reference 11. Also, in Table 2 the values of $S/2Sy < 1$ do not mean that the pipe could not withstand higher dynamic loads; rather, no attempt was made to apply higher dynamic loads.

Thus, the staff finds that the dynamic test results clearly demonstrate that with certain limitations discussed in Sections 8.1.3, 8.1.5, and 8.1.7, Code Equation (9) with a stress limit of $2Sy$, using 5% damping, provides assurance that piping functional capability will be maintained.

8.1.3 Steady-State Stresses

Weight stresses should be considered as design conditions. The Code limit on Equation (9) for design conditions is $1.5Sx$, where $Sx = Sm$ for Class 1 piping, $Sx = Sh$ for Class 2 piping. In the bounding case in which $P = 0$, $Sm = Sh = 0.9Sy$ (austenitic steel at 650°F), the

---

*N/A = not applicable; results were obtained from Section 5 of this report.

allowable moment due to weight using Equation (9), for $1.5Sx = 1.35Sy$, and for straight pipe ($B_2 = 1.0$), is

$$M = 1.35 \, ZSy$$

Equation (2) gives

$$M_L = (4/\pi)ZSy = 1.27 \, ZSy$$

Thus, in this bounding case, the moment due to weight is about equal to the theoretical limit moment.

At Level D, the stress due to combinations of weight plus dynamic loads is limited to $2Sy$. Thus, at Level D, there is a spectrum of allowable combinations ranging from $Sw = 0, Sd = 2.0Sy$, to $Sw = 1.35Sy, Sd = 0.65Sy$, where Sw = weight stress, Sd = dynamic stress.

The test evaluations clearly indicate that the combination of $Sw = 0, Sd = 2.0Sy$ maintains functional capability.

Unfortunately, no tests are available that show that the combination of $Sw = 1.35Sy, Sd = 0.65Sy$ maintains functional capability. That is, if a straight pipe were loaded to its limit load by weight and then subjected to a dynamic stress of $+/-0.65Sy$, would functional capability be maintained?

In the absence of relevant test data, the staff recommends that steady-state stresses be limited to $0.25Sy$. Its judgment is based mainly on Reference 11 Tests 30 and 37. In these tests, the combinations were

Test 30 $S/2Sy$ =   5.2 (5% damping), $Sw/Sy = 0.32$, no collapse

Test 37/4 $S/2Sy$ =   2.8 (5% damping), $Sw/Sy = 0.32$, no collapse

Test 37/5 $S/2Sy$ =   5.5 (5% damping), $Sw/Sy = 0.32$, collapse

Using Code Equation (9) Level D limit of $2Sy$,

$$S/2Sy = (2-0.32)/2 = 0.84$$

Comparing this value with the corresponding value from Test 37, Run 4, indicates that permitting Sw/Sy up to 0.25 is adequate to ensure maintenance of functional capability. The value of 0.25 was deliberately chosen to be a bit less than that in Test 37.

A conceptually more direct method of controlling steady-state stresses might be to introduce an Equation (9a), which would directly, and independently of Code Equation (9), limit steady-state stresses to $0.25Sy$. However, the staff's present goal is to be able to say that meeting

Code Equation (9) Level D limits also ensures functional capability.

The staff's recommended limit on steady-state stresses of $0.25Sy$ is not deemed to be onerous if the steady-state stresses are due to weight. Typically, weight stresses do not exceed about 3 ksi. Some examples of the $0.25Sy$ limit are

Material	Temp., °F	Sy, ksi	$0.25Sy$, ksi
A106–B	100	35	8.75
A106–B	650	25.4	6.35
Type 304	100	30	7.5
Type 304	650	17.9	4.48

However, if the steady-state stress is due to the steady-state thrust of a relief-valve discharge, then the staff's Sw limit of $0.25Sy$ might be restrictive.

8.1.4 Pressure Stresses

Code Equation (9) includes the term [see Equation (1) herein] $B_1PD_o/2t$. For most components, $P > 0$, this term reduces the allowable combinations of Sw and Sd.

The staff's test data evaluations did not indicate any adverse effect of $P > 0$ on functional capability. Indeed, for elbows, $P > 0$ tended to increase the moment capacity. This aspect is partially recognized in the Code by

$$B_1 = -0.1 + 0.4h \text{ but not } < 0 \text{ nor } > 0.5$$

Thus, for elbows with $h < 1/4$, the pressure term becomes zero. But for other components, and elbows with $h > 1/4$, the pressure term, $P > 0$, tends to add to the margin for assurance of functional capability.

However, there is a potential for external pressure to jeopardize functional capability. An external pressure might arise for piping inside the containment when the containment is pressurized under accident conditions. The staff's D_o/t limit (see Section 8.1.5) partially addresses this concern. However, the staff's recommendations include a restriction that external pressure must not exceed internal pressure, as a reminder that this special condition might need to be considered.

8.1.5 D_o/t Limit

The available test data are mostly for components with $D_o/t < 50$, for example, 6 NPS, Sch. 10, $D_o/t = 6.625/0.134 = 49.4$. Three Beaney tests (Ref. 10) were on straight pipe with $D_o/t = 103/1.5 = 69$, but incipient buckling occurred in one of these three tests.

Thus, the staff deems it prudent to limit its recommendations for functional capability evaluation to components with $D_o/t < 50$. The Code also applies this limit to applicability of B-indices.

8.1.6 Future Changes in B-indices

The staff's recommendations are based on B-indices as given in the present Code (Ref. 1). Code committees constantly review newly developed data relevant to stress indices and, sometimes, these reviews lead to reducing the magnitude of stress indices. However, the Code committees are interested in pressure boundary integrity, not necessarily functional capability. Thus, it becomes incumbent on the NRC staff to review any future Code changes in B-indices from the standpoint of their effect on functional capability.

8.1.7 Future Changes to Code Equation (9) Stress Limits

Code committees have been reviewing from the standpoint of pressure boundary integrity the same sets of test data reviewed in this report from the standpoint of functional capability. It is possible that the Code Equation (9) Level D limit of $2Sy$ might be increased to $4Sy$.

It would be highly desirable that, if the Level D limit were increased to $4Sy$, it could be demonstrated that meeting the Code would also ensure functional capability.

As stated in Section 8.1.3,

Test 30 $S/2Sy = 5.2$ (5% damping), $Sw/Sy = 0.32$, no collapse

Test 37/4 $S/2Sy = 2.8$ (5% damping), $Sw/Sy = 0.32$, no collapse

Test 37/5 $S/2Sy = 5.5$ (5% damping), $Sw/Sy = 0.32$, collapse

Using the Code Equation (9) Level D limit of $4Sy$,

$$S/2Sy = (4-0.32)/2 = 1.84$$

Comparing this allowable value (1.84) with Test 37, Run 4, $S/2Sy = 2.8$, no collapse, indicates that permitting Sw/Sy up to 0.25 is adequate to ensure maintenance of functional capability, even if the Level D limit on Code Equation (9) is increased to $4Sy$.

However, the boundary between static loading and dynamic loading is not well-defined. Use of a Code Equation (9) limit of $4Sy$ can only be defended by the available test data for rapidly reversing dynamic loads. For Component Test 37, discussed in Section 4.6, and the HEDL test, discussed in Section 6.1, the dominant response frequencies were about 2 Hz. Both of these tests resulted in an incipient threat to functional capability. Thus, the staff believes that it is prudent to restrict a Code Equation (9) limit of $4Sy$ to piping systems for which the elastic response spectrum analysis indicates that the response stress contribution at 2 Hz and less is not more than Sy.

8.2 Other Dynamic Loads

Section 7 contains a brief discussion of other dynamic loads. The staff concludes that it is appropriate to include fluid-hammer pressure wave loads in the category of reversing dynamic loads. Those dynamic loads that are not clearly in the category of reversing dynamic loads, and combinations of reversing with nonreversing dynamic loads, will require special consideration. Some suggestions are included in Section 7.

9 CONCLUSIONS

9.1 Functional Capability Assurance, Present Code Requirements

The staff concludes that piping functional capability is ensured by meeting the present Code (Ref. 1) requirements, provided

(1) Dynamic loads are reversing. This includes loads due to earthquakes, building-filtered loads such as those due to vibration of buildings caused by relief-valve actuation in boiling-water reactors, and pressure wave loads (not slug-flow fluid hammer).

(2) Dynamic moments are calculated using an elastic response spectrum analysis with +/-15% peak broadening and with not more than 5% damping.

(3) Steady-state (e.g., weight) stresses do not exceed $0.25S_y$.

(4) D_o/t does not exceed 50.

(5) External pressure does not exceed internal pressure.

9.2 Functional Capability Assurance, Future Code Requirements

Until such time as Code changes are made, the staff can make no specific conclusions concerning such changes.

If the Code Equation (9) Level D limit is increased to, for example, $4S_y$, the staff concludes that in addition to restrictions (1) through (5) in the previous section, an additional restriction would be needed; that is, the elastic response spectrum analysis must show that the response stress contribution at 2 Hz and less is not more than S_y. (See Section 8.1.7.)

Any changes in B-indices in the present Code should be reviewed to determine whether such changes would adversely affect the assurance of functional capability.

With the use of a limit greater than $2S_y$, increased vigilance would be needed to provide assurance that such components as piping supports, anchors, restraints, guides, and anchors have sufficient load capacity.

10 REFERENCES

1. American Society of Mechanical Engineers, Boiler and Pressure Vessel Code, Section III, Division 1, "Nuclear Power Plant Components," New York, 1989 Edition.

2. United States of America Standard B31.1-1967, "USA Standard Code for Pressure Power Piping," American Society of Mechanical Engineers, New York.

3. U.S. Nuclear Regulatory Commission, NUREG/CR-0261, "Evaluation of the Plastic Characteristics of Piping Products in Relation to ASME Code Criteria," Rodabaugh and Moore, July 1978.

4. ---, NUREG-0800, "Standard Review Plan for the Review of Safety Analysis Reports for Nuclear Power Plants," Section 3.9.3, "ASME Code Class 1, 2, and 3 Components, Component Supports, and Core Support Structures," Rev. 1, July 1981.

5. General Electric Company, Nuclear Energy Engineering Division, "Functional Capability Criteria for Essential Mark II Piping," NEDO-21985, San Jose, California, September 1978.

6. U. S. Nuclear Regulatory Commission, NUREG-1061, "Report of the U. S. Nuclear Regulatory Commission Piping Review Committee" (5 Volumes), Volume 2, "Evaluation of Seismic Designs—A Review of Seismic Design Requirements for Nuclear Power Plant Piping," April 1985.

7. E. M. Beaney, "Response of Tubes to Seismic Loading," TPRD/B/0605/N85, Central Electricity Generating Board, Berkeley Nuclear Laboratories, Berkeley, Gloucestershire, United Kingdom, January 1985.

8. ---, "Response of Pipes to Seismic Excitation—Effect of Pipe Diameter/ Wall Thickness Ratio and Material Properties," TPRD/B/0637/N85, Central Electricity Generating Board, Berkeley Nuclear Laboratories, Berkeley, Gloucestershire, United Kingdom, July 1985.

9. ---, "Response of Pressurized Straight Pipe to Seismic Excitation," TPRD/B/0826/R86, Central Electricity Generating Board, Berkeley Nuclear Laboratories, Berkeley, Gloucestershire, United Kingdom, February 1986.

10. ---, "Response of Stainless Steel Pipes to Seismic Excitation," TPRD/B/1051/R85, Central Electricity Generating Board, Berkeley Nuclear Laboratories, Berkeley, Gloucestershire, United Kingdom, April 1988.

11. General Electric Company, Nuclear Energy Engineering Division, "Piping and Fitting Dynamic Reliability Program," Volume 2, "Component Test Report," EPRI Contract RP 1543-15, Draft, San Jose, California, December 1989.

12. J. Spence and G. E. Findlay, "Limit Loads for Pipe Bends Under In-Plane Bending," Paper No. I-28, *Proceedings of the 2nd International Conference on Pressure Vessel Technology,* American Society of Mechanical Engineers, New York, October 1973.

13. General Electric Company, Nuclear Energy Engineering Division, "Piping and Fitting Dynamic Reliability Program," Volume 3, "System Test Report," Draft, San Jose, California, February 1990.

14. Hanford Engineering Development Laboratory, "High-Level Dynamic Testing and Analytical Correlations for a One-Inch Diameter Piping System," M. R. Lindquist, M. J. Anderson, L. K. Severud, and E. O. Weiner, HEDL-TME 85-24, Richland, Washington, February 1986.

15. G. E. Howard, B. A. Johnson, W. B. Walton, H. T. Tang, and Y. K. Tang, "Piping Extreme Dynamic Response Studies," *Proceedings of the 7th Structural Mechanics in Reactor Technology Conference,* Vol. F, August 1983.

16. U. S. Nuclear Regulatory Commission, NUREG/CR-3893, "Laboratory Studies: Dynamic Response of Prototypical Piping Systems," ANCO Engineers, Inc., August 1984.

17. ---, NUREG/CR-5023, "High-Level Seismic Response and Failure Prediction Methods for Piping," Westinghouse Hanford Co., January 1988.

18. B. Charalambus, E. Haas, and R. Mihatsch, "Comparisons of Dynamic Test Data with Results of Various Analytical Methods," *Nuclear Engineering and Design,* Vol. 96, pp. 447-462, 1986.

19. U. S. Nuclear Regulatory Commission, NUREG/CR-5757, "Verification of Piping Response Calculation of SMACS Code with Data from Seismic Testing of an In-Plant Piping System," Argonne National Laboratory, September 1991.

20. ---, NUREG/CR-5585, "The High Level Vibration Test Program," Brookhaven National Laboratory, May 1991.

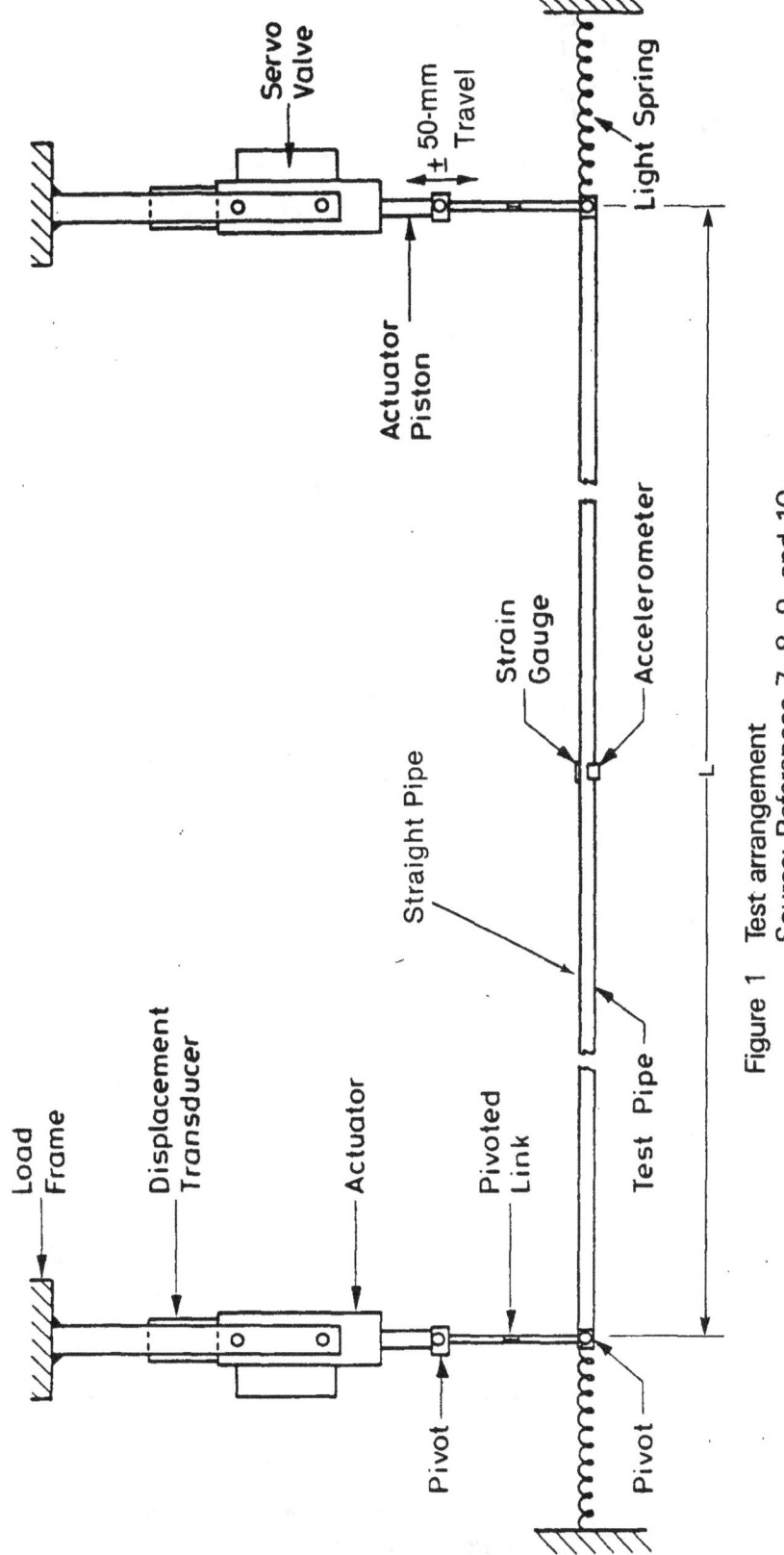

Figure 1 Test arrangement
Source: References 7, 8, 9, and 10.

33

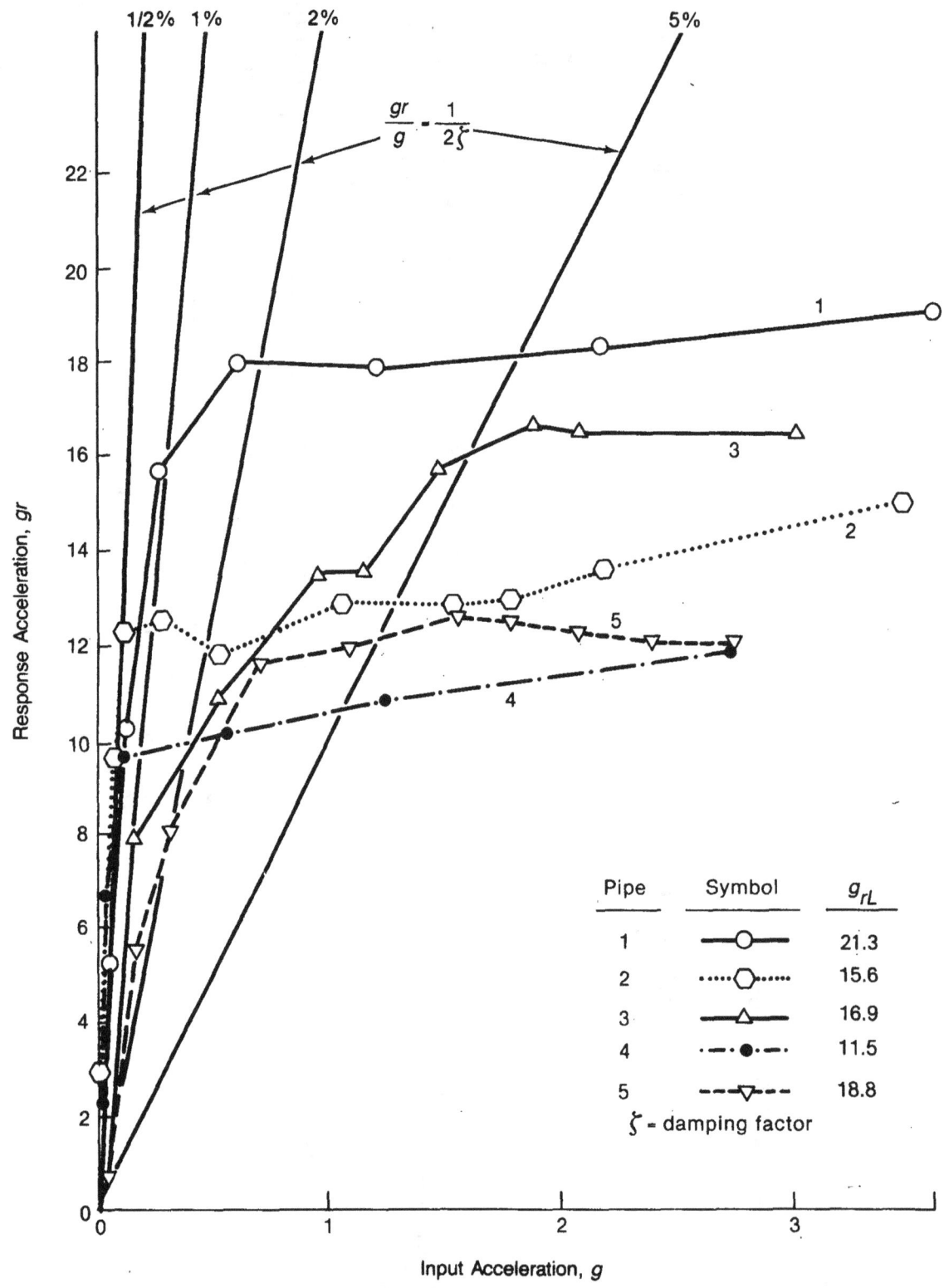

Figure 2   Response versus input acceleration
Source: Reference 8.

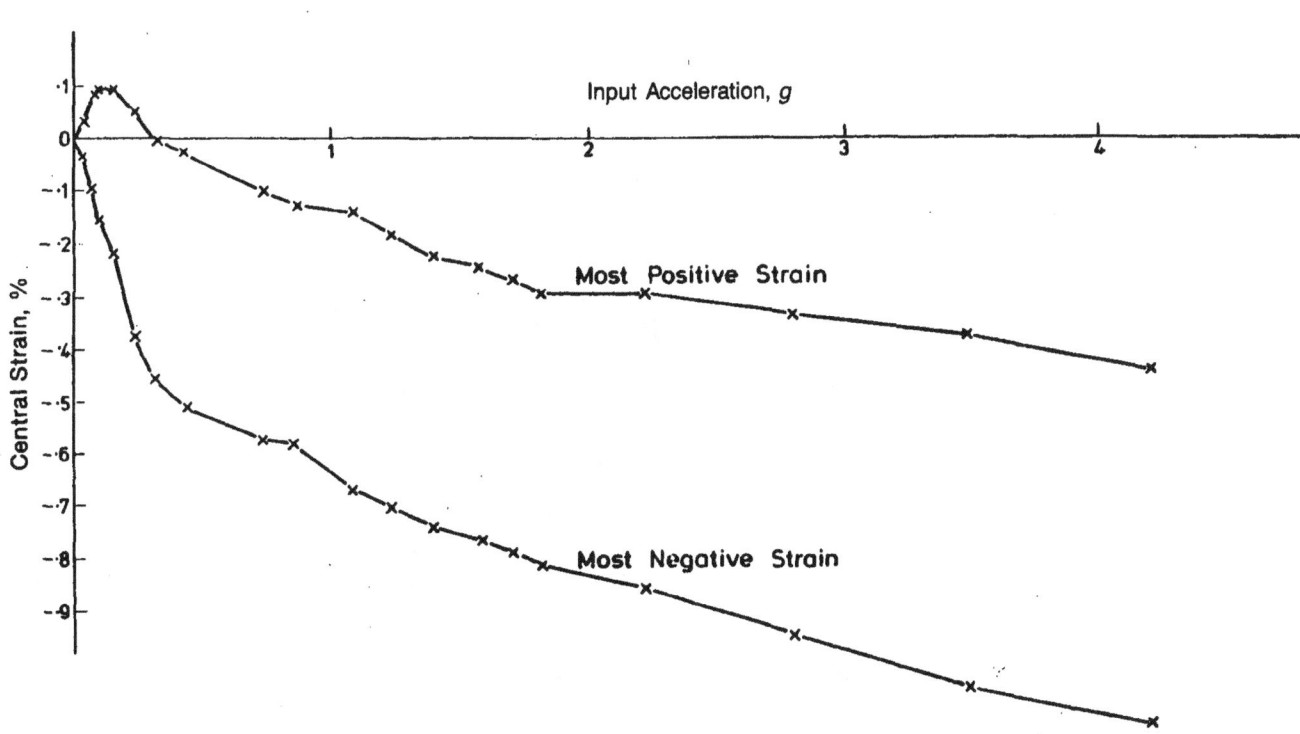

Figure 3   Strain at pipe midspan versus input acceleration
Source: Reference 7.

NUREG-1367

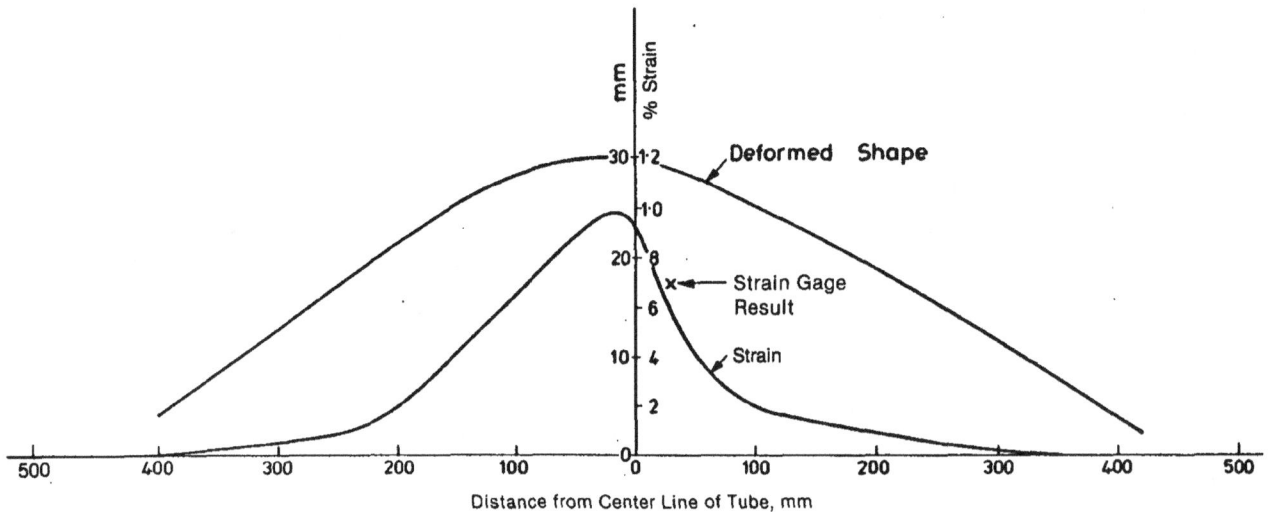

Figure 4   Deformed shape and permanent strain after tests
Source: Reference 7.

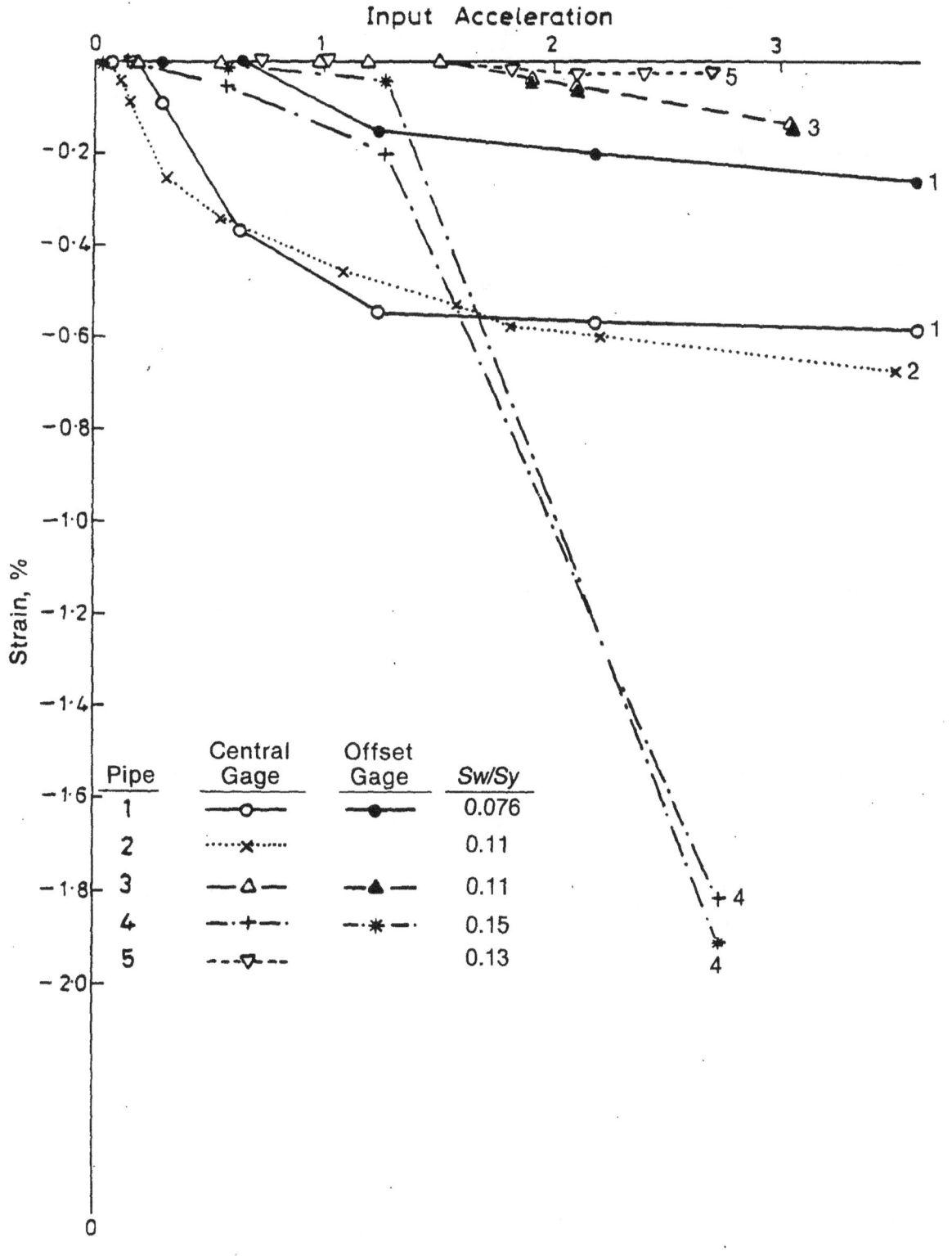

Figure 5   Mean strain versus input acceleration
Source: Reference 8.

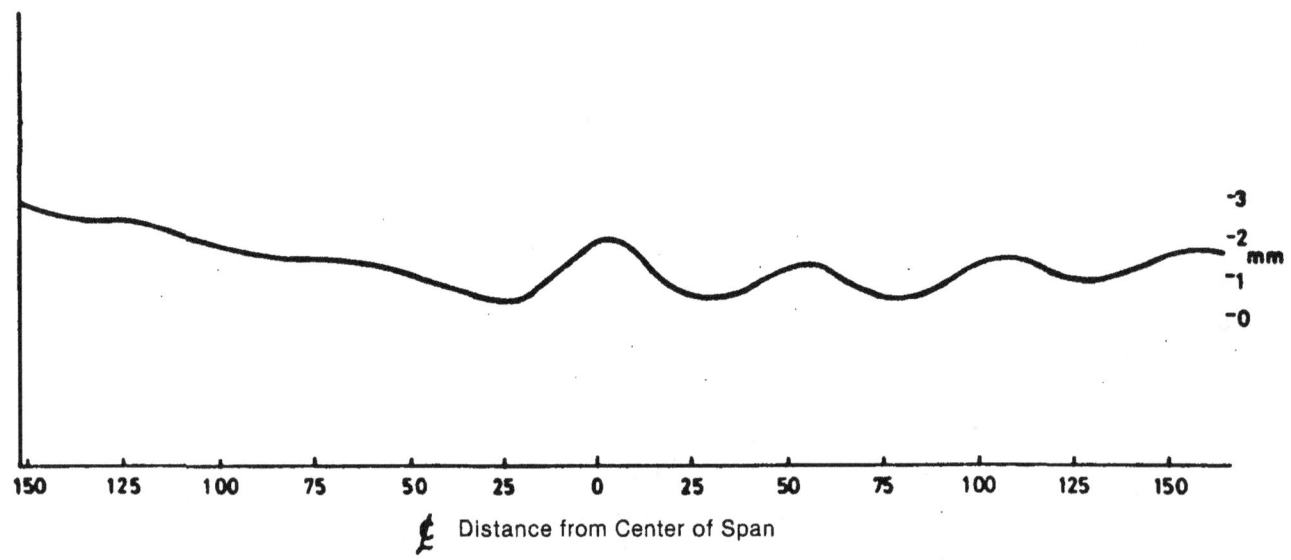

Figure 6   Deformed shape of upper surface of 103-mm pipe, Test 16
Source: Reference 10.

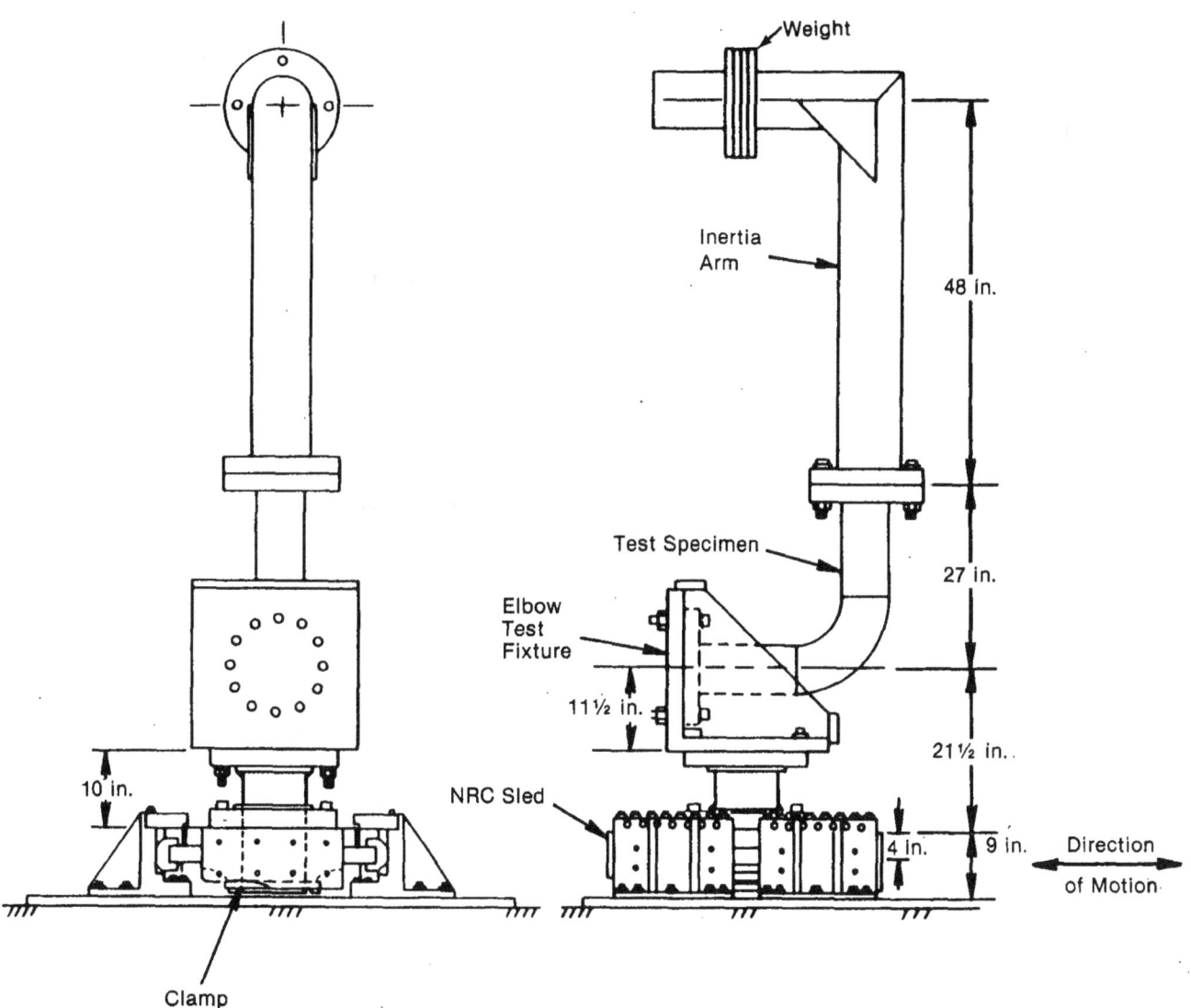

Figure 7   In-plane elbow test arrangements, Tests 1, 3-8, 13, 19, and 31
Source: Reference 11.

NUREG–1367

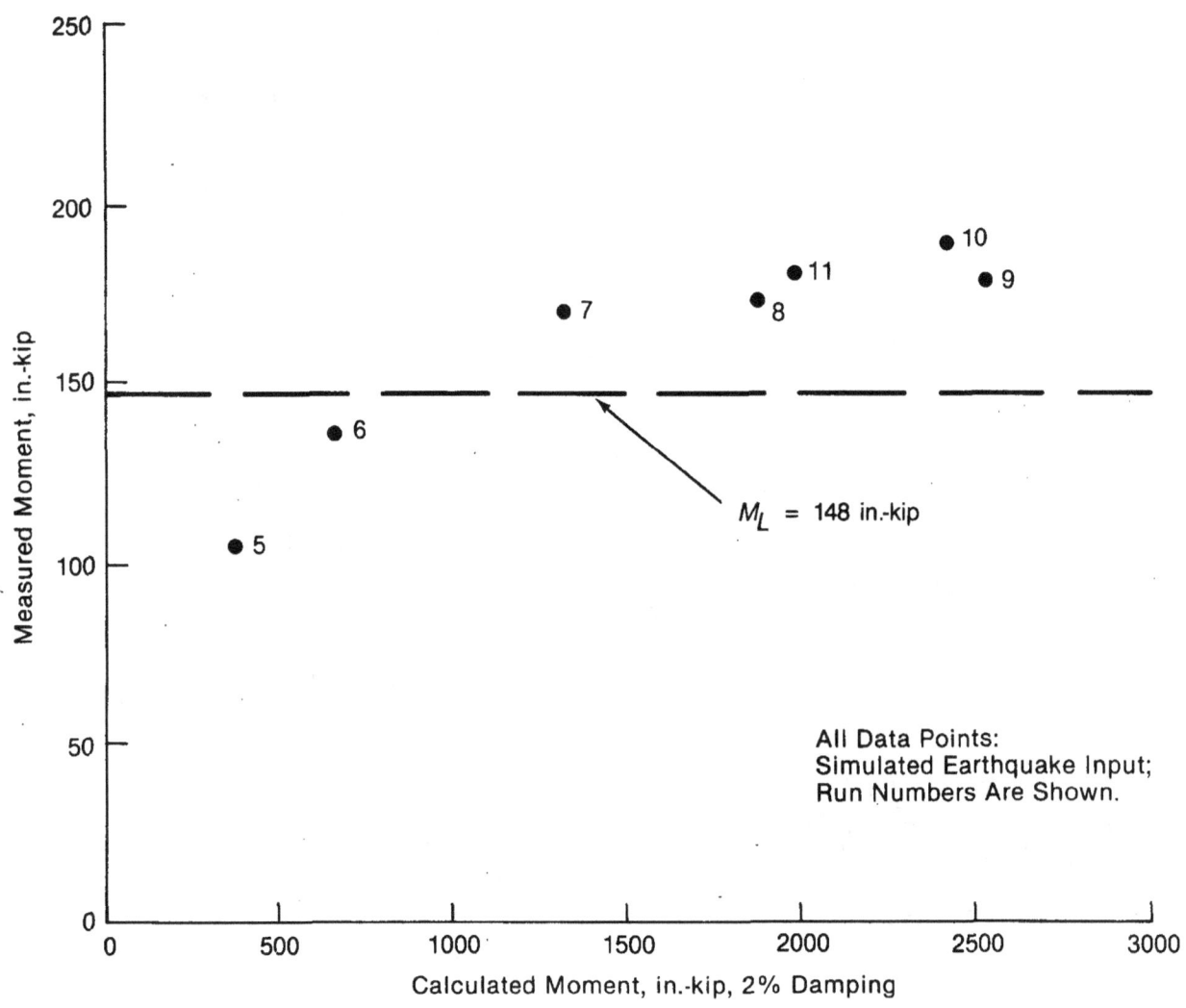

Figure 8   4 NPS, Sch. 40 stainless steel pipe, Test 15
Source: Reference 11.

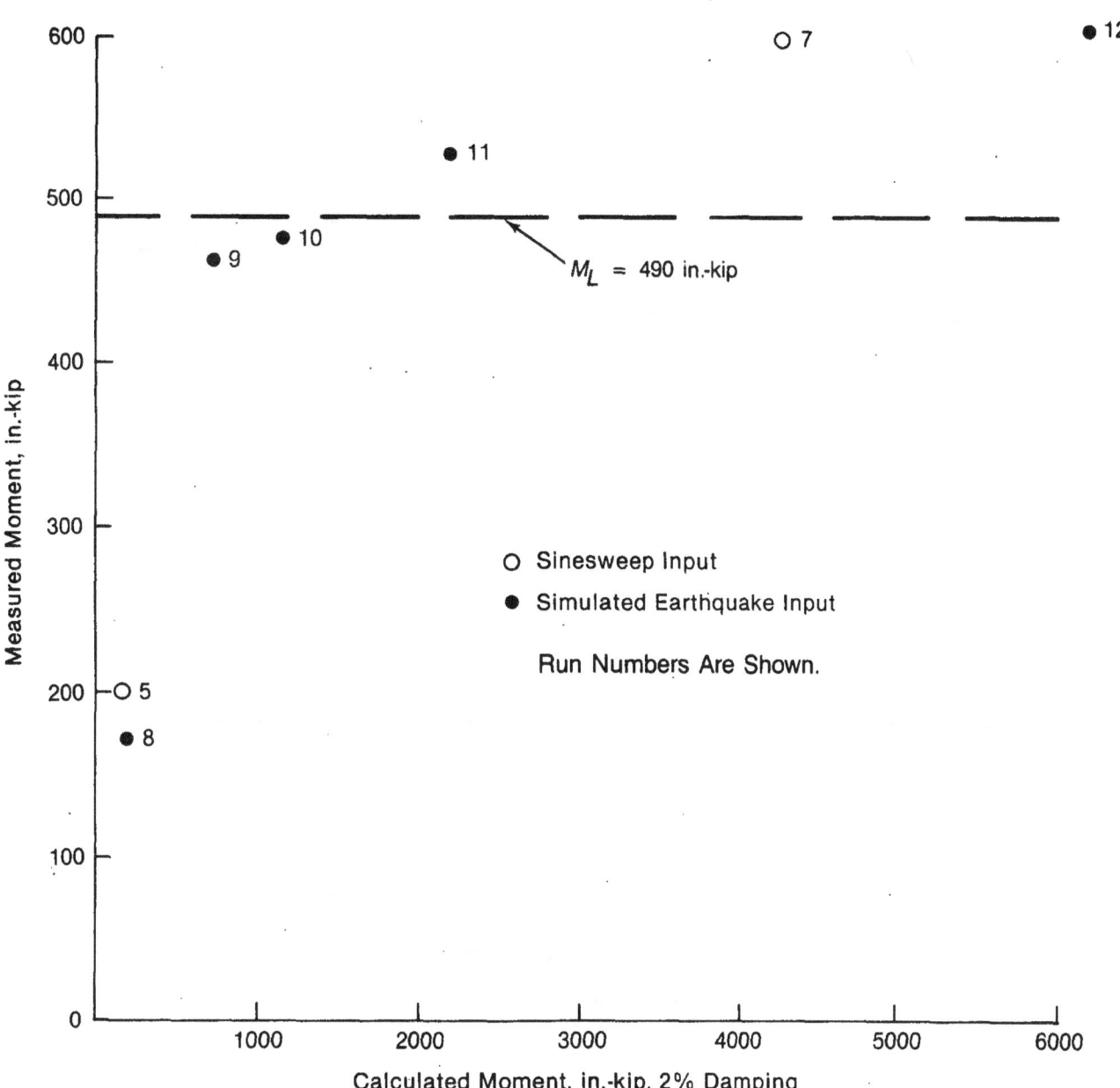

Figure 9   6 NPS, Sch. 40 carbon steel pipe, Test 34
Source: Reference 11.

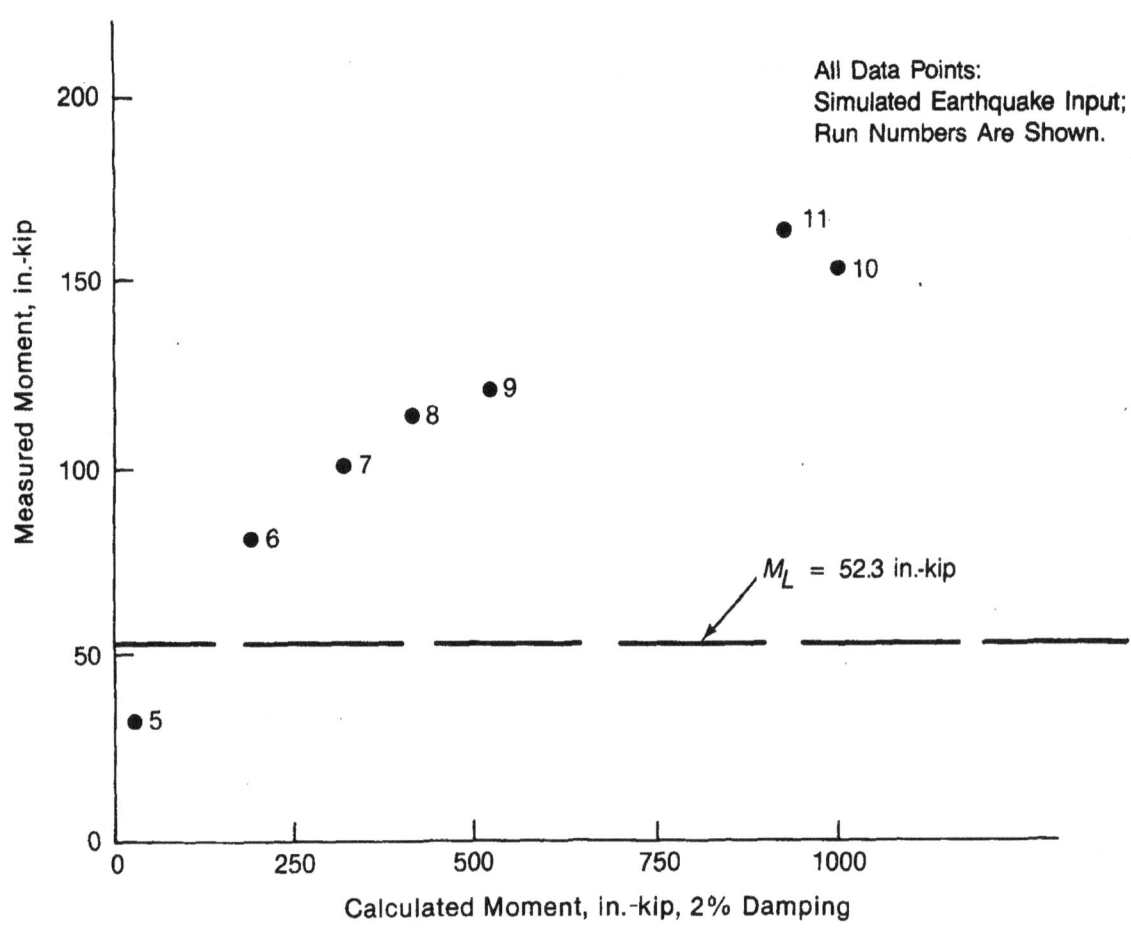

Figure 10   6 NPS, 9-in. bend radius, Sch. 10 stainless steel elbow, Test 3
Source: Reference 11.

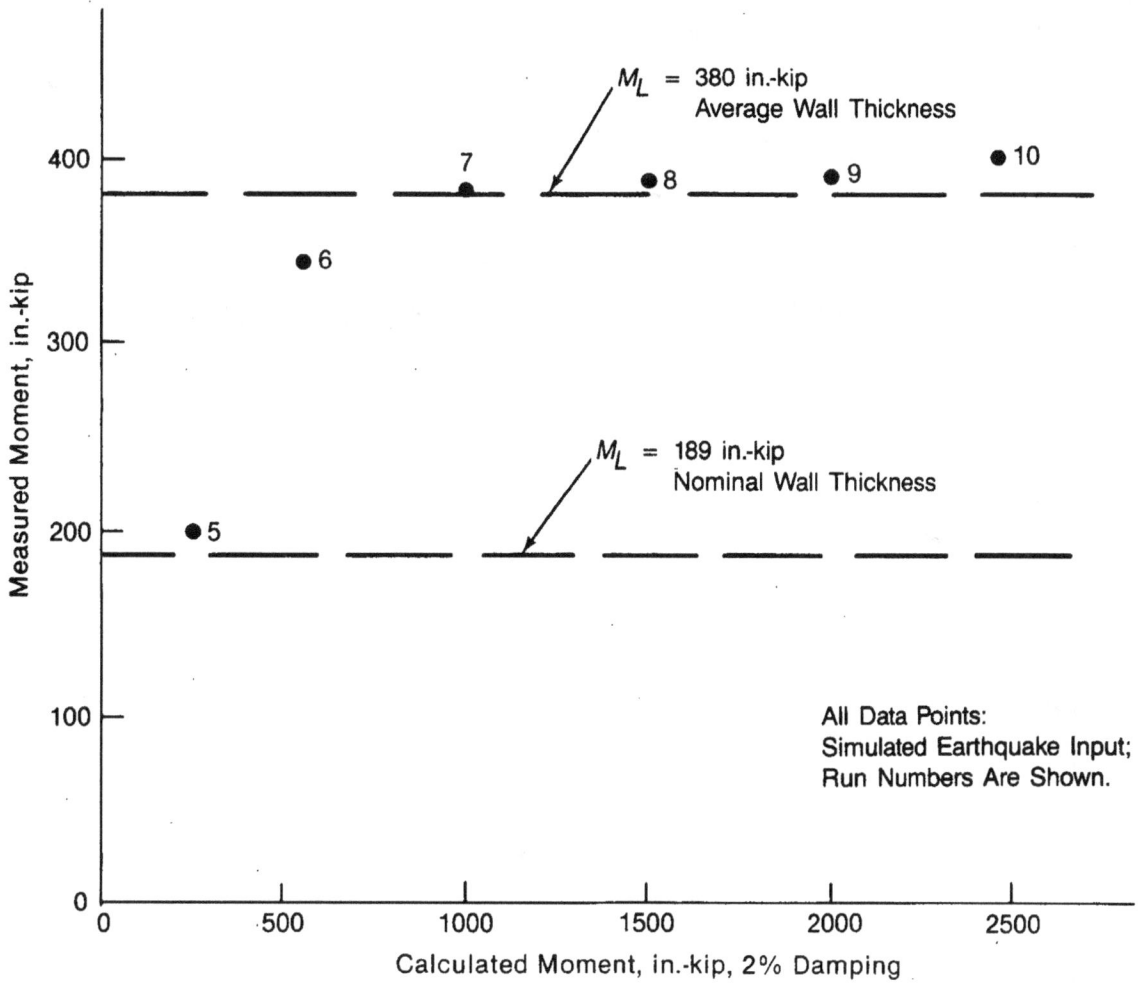

Note:
This elbow had a measured average wall thickness
of 1.52 times nominal wall thickness.

Calculated moments are based on nominal wall thickness.

Figure 11   6 NPS, 6-in. bend radius, Sch. 40 carbon steel elbow, Test 13
           Source: Reference 11.

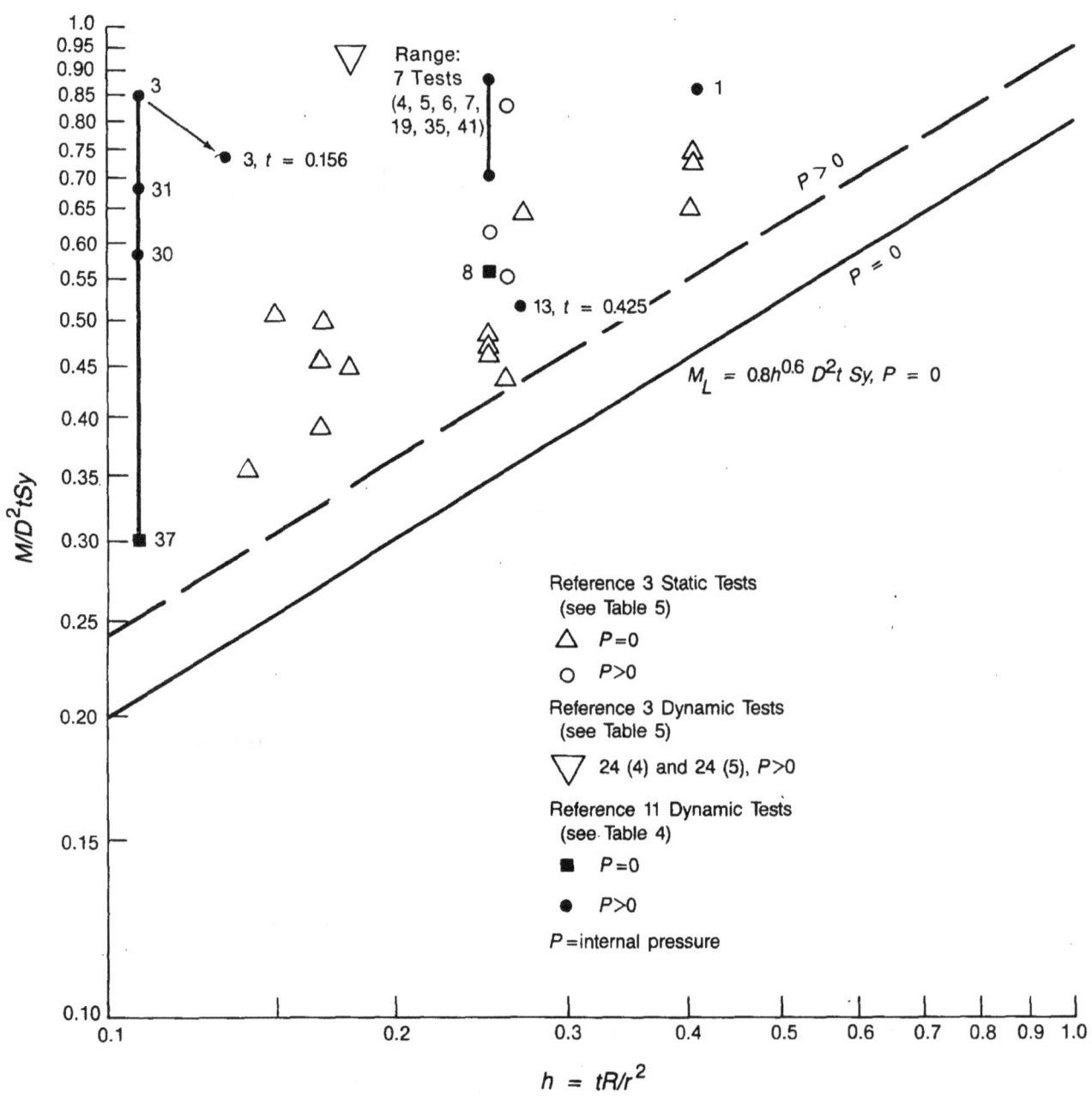

Note: Number next to symbol is test number.

Figure 12   Elbows: Static in-plane closing moment capacity and dynamic in-plane
moment capacity tests
Source: References 3 and 11.

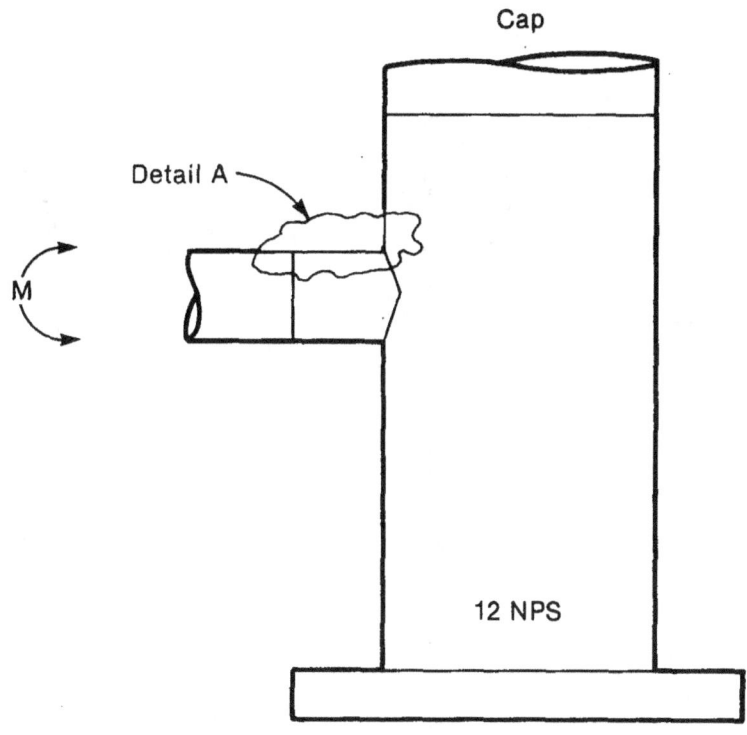

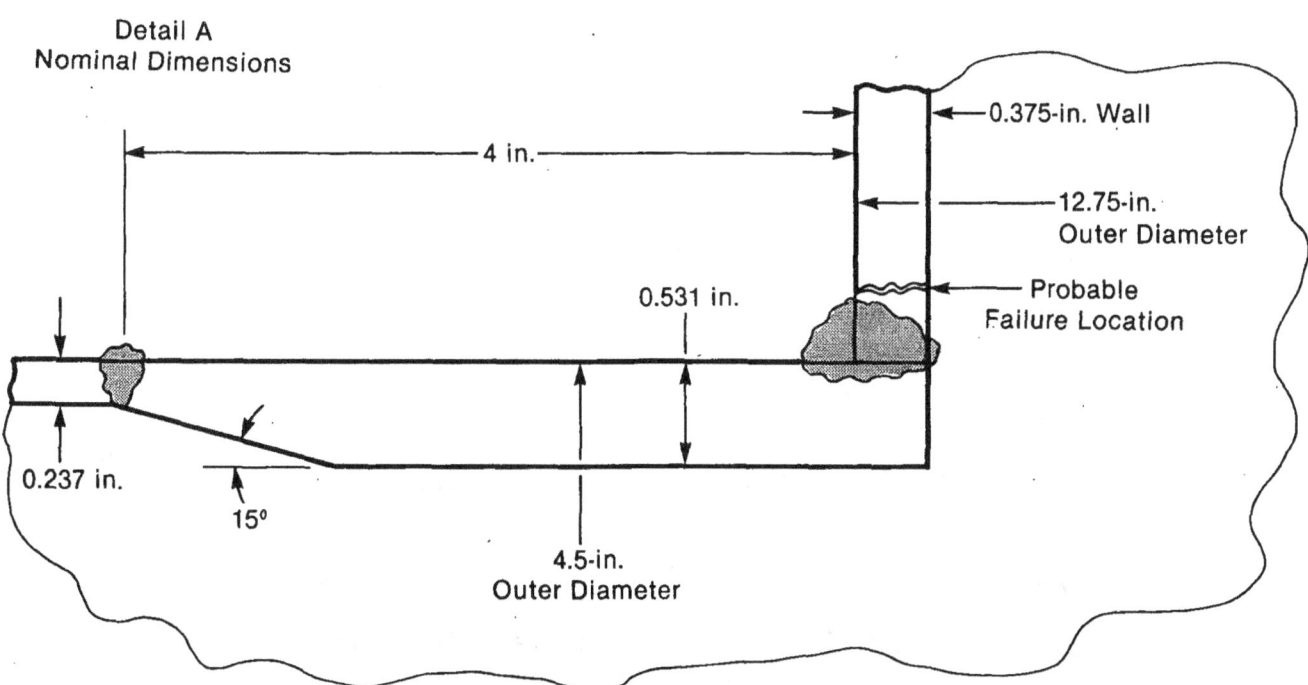

Figure 13   Test 20 configuration: 4 NPS nozzle in 12 NPS vessel
Source: Reference 11.

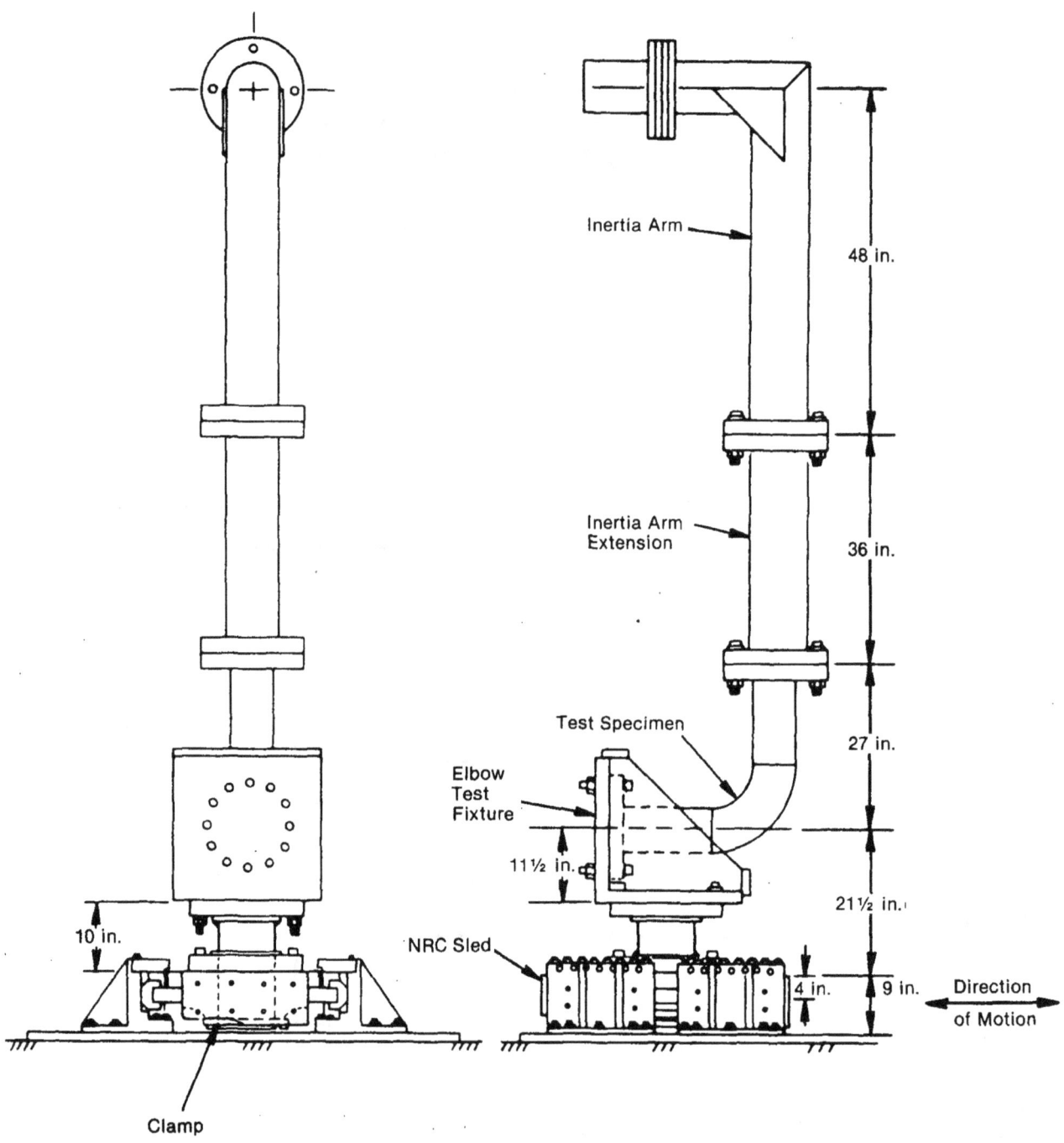

Figure 14   Test arrangement, Tests 30 and 37
Source: Reference 11.

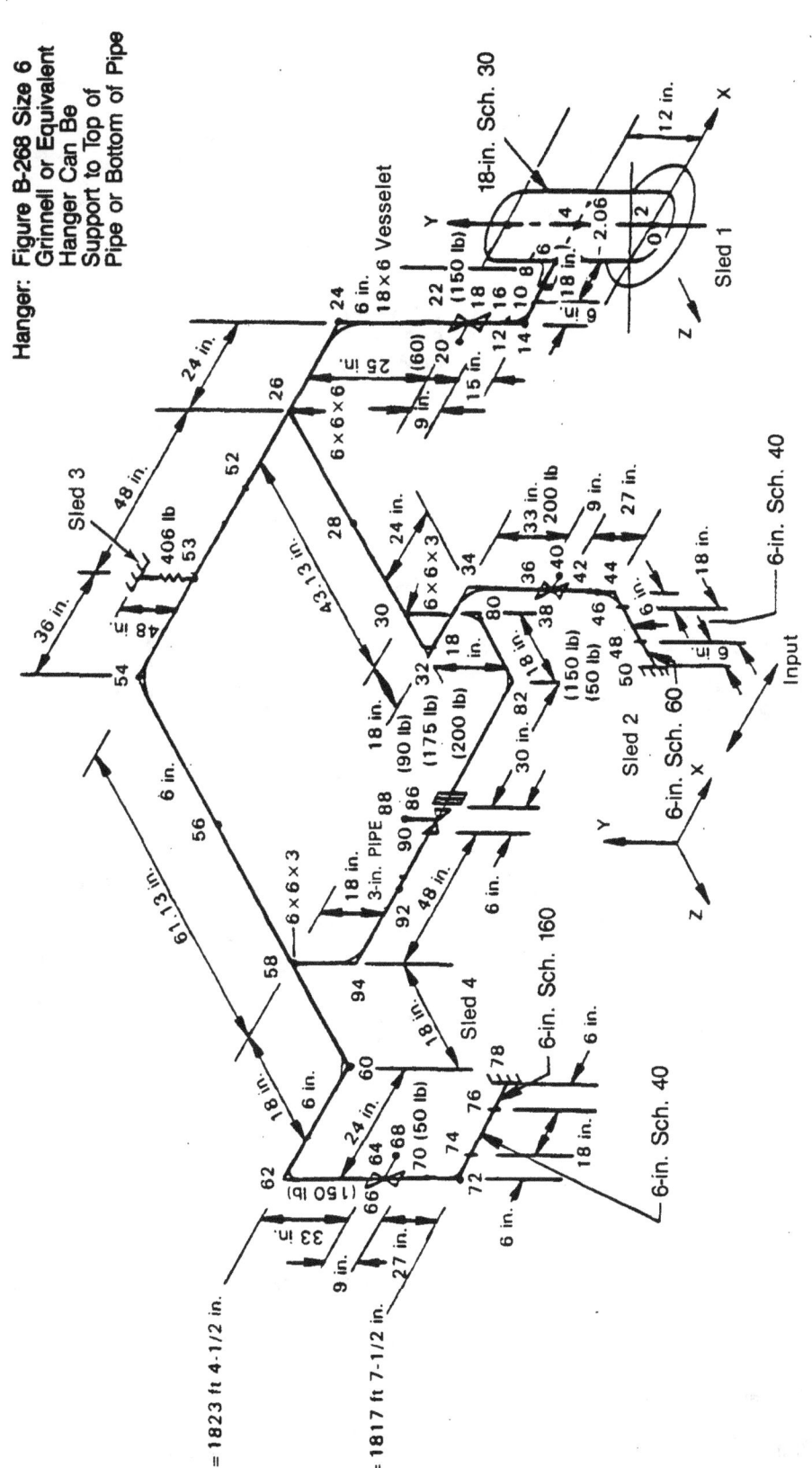

Figure 15   Piping System 1 configuration [material: carbon steel (A106-B)]
Source: Reference 13.

Note: Valves 18, 38, and 66 are simulated by lump weight.
Valve 86 is a motor-operated valve.

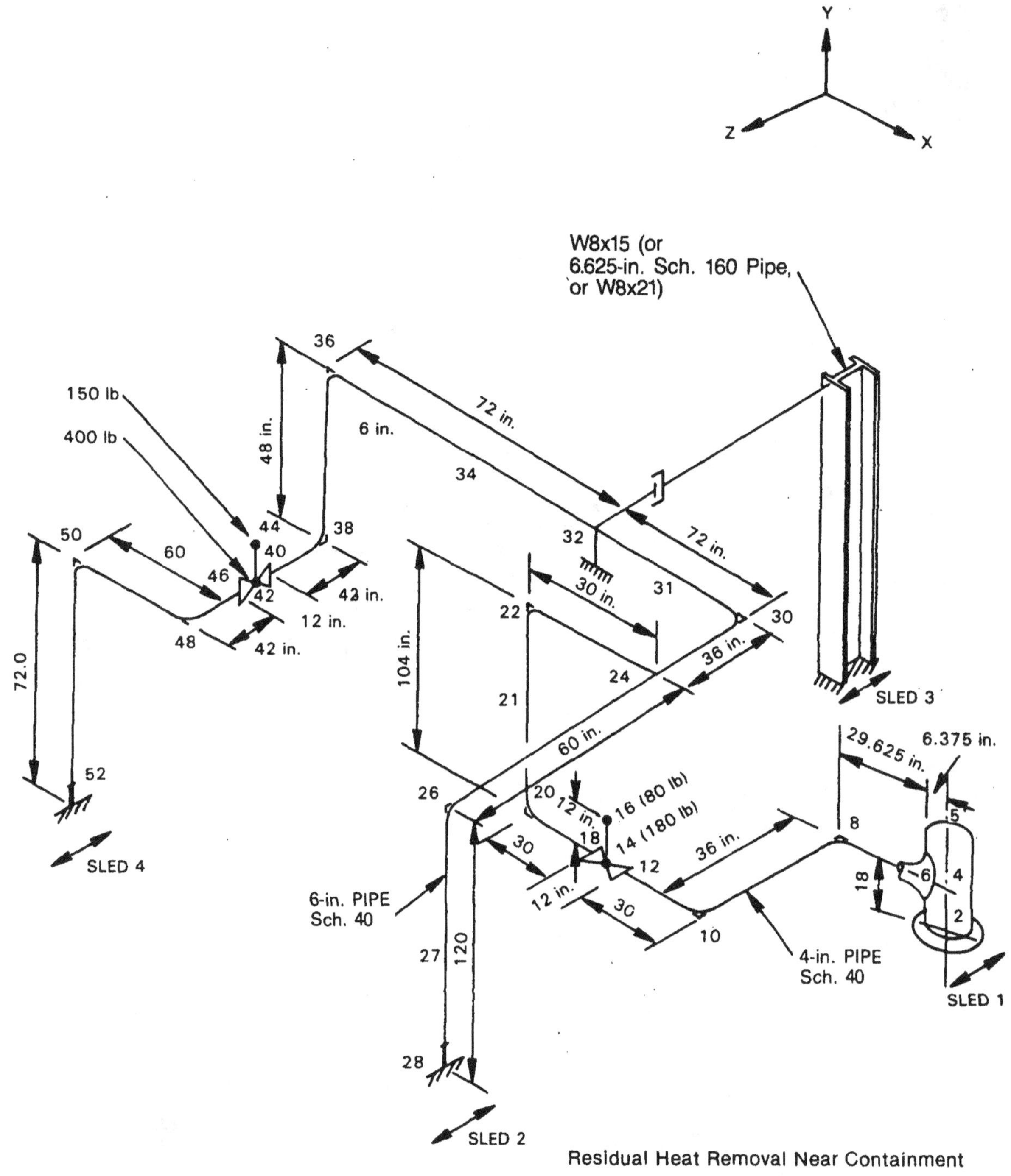

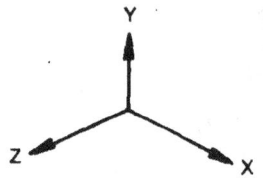

Y
Z ← → X

W8x15 (or
6.625-in. Sch. 160 Pipe,
or W8x21)

36

150 lb

400 lb

48 in.

6 in.

72 in.

34

50
60
46
44   38
40
42
48
42 in.
12 in.
42 in.

32
30 in.
31

72 in.

104 in.

22
21
24
60 in.
36 in.
30

SLED 3

72.0

52

SLED 4

26
20
18
16 (80 lb)
14 (180 lb)
12
30
36 in.
10

29.625 in.
6.375 in.

8
18
6   4
5
2

6-in. PIPE
Sch. 40

30
12 in.
30

27   120

4-in. PIPE
Sch. 40

SLED 1

28

SLED 2

Residual Heat Removal Near Containment

Note: Valves 42 and 14 are simulated by weight.

Figure 16   Piping System 2 configuration [material: stainless steel (Type 316)]
Source: Reference 13.

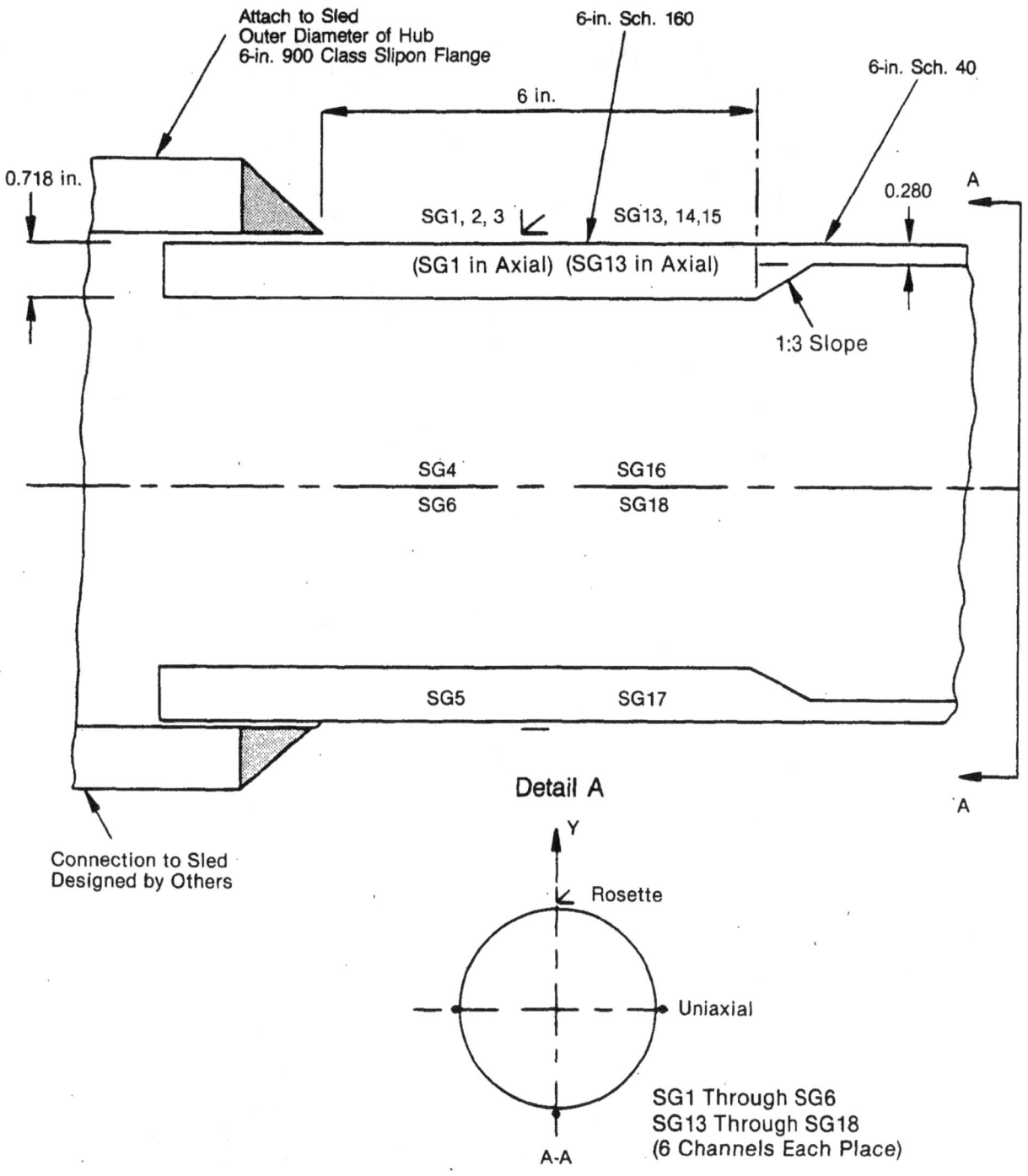

Detail A

SG1 Through SG6
SG13 Through SG18
(6 Channels Each Place)

Typical for Two Places

Note: SG = strain gauge

Figure 17   Load measurement device at Sleds 2 and 4
Source: Reference 13.

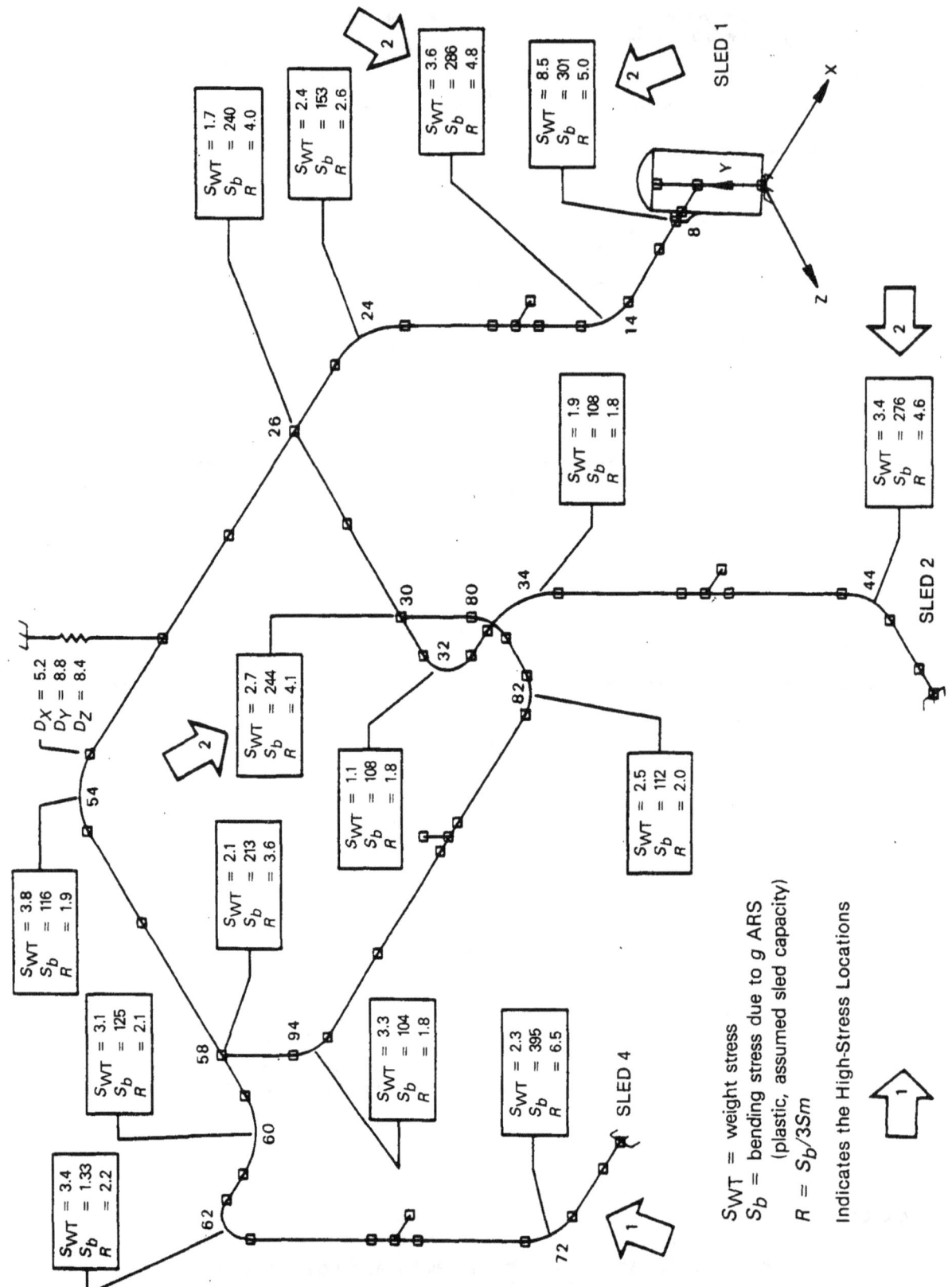

Figure 18 Piping System 1, weight stresses, S_{WT}
Source: Reference 13.

S_{WT} = weight stress
S_b = bending stress due to g ARS
     (plastic, assumed sled capacity)
$R = S_b/3S_m$

Indicates the High-Stress Locations

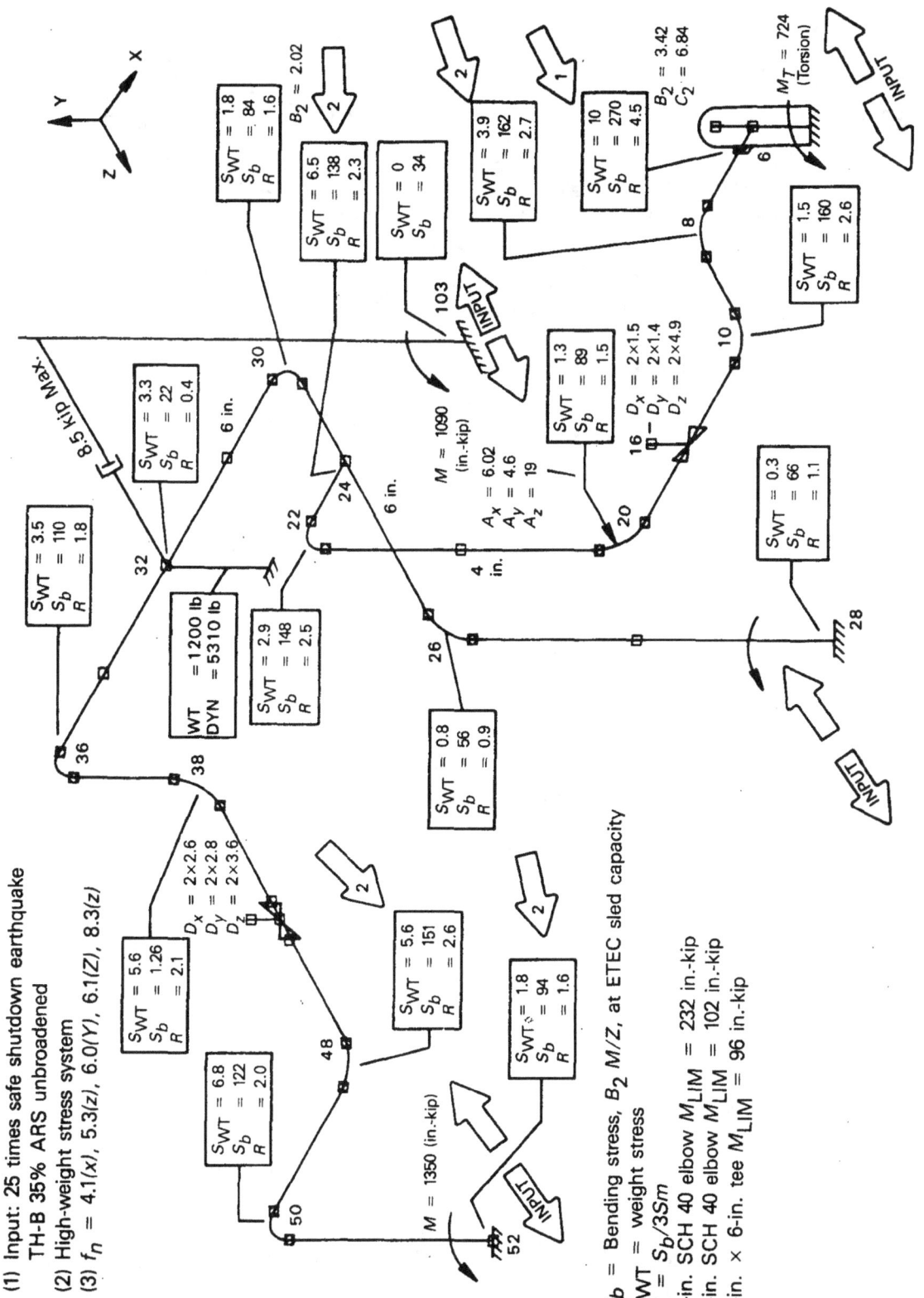

Figure 19 Piping System 2, weight stresses, S_{WT}
Source: Reference 13.

(1) Input: 25 times safe shutdown earthquake
    TH-B 35% ARS unbroadened
(2) High-weight stress system
(3) $f_n = 4.1(x), 5.3(z), 6.0(Y), 6.1(Z), 8.3(z)$

S_b = Bending stress, $B_2 M/Z$, at ETEC sled capacity
S_{WT} = weight stress
$R = S_b/3S_m$
6-in. SCH 40 elbow M_{LIM} = 232 in.-kip
4-in. SCH 40 elbow M_{LIM} = 102 in.-kip
4-in. × 6-in. tee M_{LIM} = 96 in.-kip

51

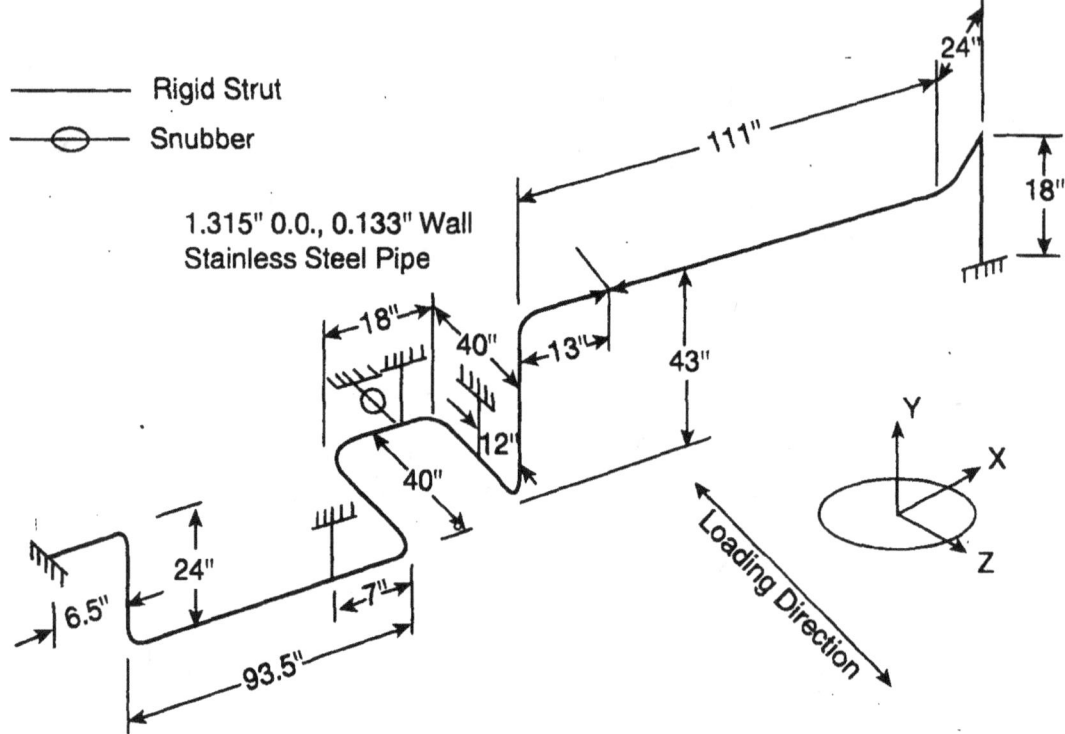

(a) 1-in.-Diameter Pipe Loop, Modified Four-Support Configuration

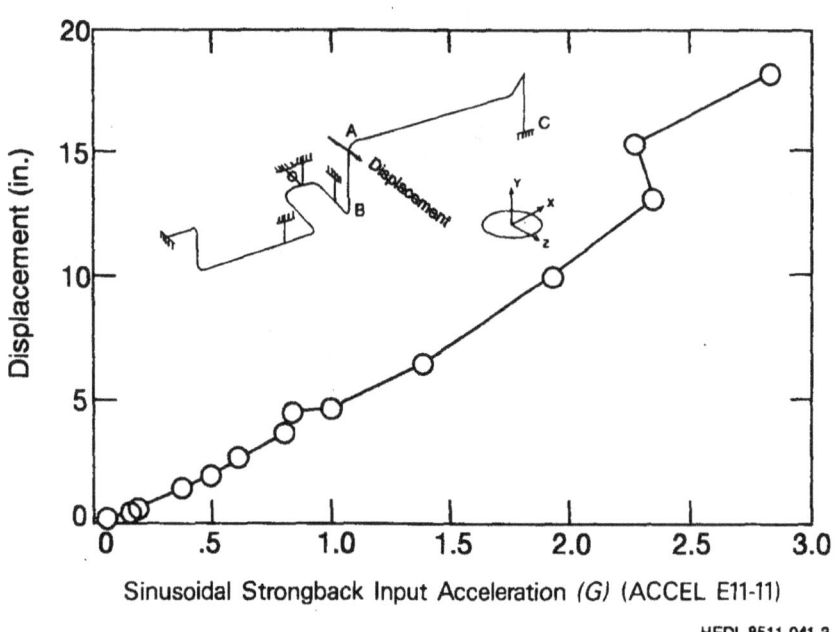

HEDL 8511-041.3

(b) High-Level Sinusoidal Test, Permanent Displacement at Upper Elbow

Figure 20   Hanford Engineering Development Laboratory piping system
Source: Reference 14.

Table 1 Beaney (Refs. 7, 8, 9, and 10) Straight Pipe Tests: Materials, Yield Strengths, Dimensions, Sinusoidal Input Test Frequencies, Pressures, and Test Planes

Ref.	Test (a)	Mtl. (b)	S_y, MN/m^2 (c)	D_o, mm (d)	t, mm (e)	L, mm (f)	f, Hz (g)	P, MN/m^2 (h)	Test Plane (i)
7	(j)	CS	209	25.4	2.64	3530	5	0.00	V
8	1	CS	298	25.4	0.91	3739	5	0.00	V
	2	CS	219	25.4	2.64	3617	5	0.00	V
	3	CS	248	25.4	6.35	3386	5	0.00	V
	4	CS	161	34.1	4.06	4165	5	0.00	V
	5	CS	223	51.8	4.47	5664	5	0.00	V
9	1	CS	162	34.14	4.06	4166	5	29.2	V
	2	CS	162	34.14	4.06	4166	5	29.2	V
	3	CS	162	34.14	4.06	4166	5	29.2	V
	4	CS	298	25.4	0.91	3734	5	14.3	V
	5	CS	162	34.14	4.06	2946	9	29.2	V
	6	CS	298	25.4	0.91	2642	9	14.3	V
	7	CS	298	25.4	0.91	2642	9	14.3	V
	8	CS	162	34.14	4.06	2946	9	29.2	V
10	1	CS	162	34.1	4.06	2946	9.7	0.00	H
	2	CS	162	34.1	4.06	2946	9.4	29.5	H
	3	CS	162	34.1	4.06	4166	4.8	0.00	H
	4	CS	162	34.1	4.06	4166	4.7	29.5	H
	5	CS	195	25.4	2.64	3556	4.8	0.00	H
	6	CS	195	25.4	2.64	3556	4.6	31.0	H
	7	SS	247	25.4	2.64	3556	4.7	0.00	H
	8	SS	247	25.4	2.64	3556	4.6	38.2	H
	9	SS	247	25.4	2.64	2891	6.7	38.2	H
	10	SS	247	25.4	2.64	2946	6.5	57.4	H
	11	SS	261	78	1.5	6121	5.2	0.00	H
	12	SS	261	78	1.5	3988	7.5	0.00	H
	13	SS	261	78	1.5	3988	7.5	10.3	H
	14	SS	335	103	1.5	6039	7.3	0.00	V
	15	SS	335	103	1.5	5490	5.1	0.00	V
	16	SS	335	103	1.5	5490	5.0	7.2	V

(a) Test identification according to references.
(b) CS = carbon steel; SS = Type 316 stainless steel.
(c) S_y = yield strength of pipe material (from references).
(d) D_o = pipe outside diameter.
(e) t = pipe wall thickness.
(f) L = pipe span length.
(g) f = sinusoidal input test frequency.
(h) P = internal pressure in pipe while being tested.
(i) V = dynamic loading in vertical plane; H = dynamic loading in horizontal plane.
(j) Straight pipe test of Reference 7.

Table 2 Beaney (Refs. 7, 8, 9, and 10) Straight Pipe Test Results Evaluated
in Relation to Elastic Analysis, 2% or 5% Damping

Ref.	Test (a)	f, Hz (b)	$M/_{gr}$ (c)	g_{mi} (d)	S_y, ksi (e)	$S/2S_y$ $\zeta = 0.02$ (f)	$S/2S_y$ $\zeta = 0.05$ (f)	S_w/S_y (g)	S_p/S_y (h)
7	(i)	5	178.4	4.2	30.3	5.19	2.08	0.11	0.00
8	1	5	67.5	3.6	43.2	2.78	1.11	0.076	0.00
	2	5	170.0	3.5	31.8	3.93	1.57	0.11	0.00
	3	5	299.9	3.0	36.0	3.40	1.36	0.11	0.00
	4	5	455.4	2.7	23.4	4.18	1.67	0.15	0.00
	5	5	1057	2.7	32.3	2.48	0.99	0.13	0.00
9	1	5	456.9	1.8	23.5	2.77	1.11	0.15	0.67
	2	5	456.9	(j)	23.5	---	---	0.15	0.67
	3	5	456.9	3.6	23.5	5.54	2.22	0.15	0.67
	4	5	67.7	0.8	43.2	0.62	0.25	0.076	0.65
	5	9	282.0	5.0	23.5	4.75	1.90	0.076	0.67
	6	9	41.7	2.1	43.2	1.00	0.40	0.038	0.65
	7	9	41.7	2.6	43.2	1.24	0.50	0.038	0.65
	8	9	282.0	4.8	23.5	4.56	1.82	0.076	0.67
10	1	9.7	241.8	5.7	23.5	4.66	1.86	0.076	0.00
	2	9.4	257.5	5.3	23.5	4.61	1.84	0.076	0.67
	3	4.8	493.9	2.8	23.5	4.67	1.87	0.153	0.00
	4	4.7	515.1	2.6	23.5	4.52	1.81	0.153	0.67
	5	4.8	190.8	3.3	28.3	4.67	1.87	0.121	0.00
	6	4.6	207.7	2.7	28.3	4.16	1.67	0.121	0.69
	7	4.7	199.0	2.8	35.8	3.27	1.31	0.095	0.00
	8	4.6	207.7	1.4	35.8	1.70	0.68	0.095	0.67
	9	6.7	148.2	4.4	35.8	3.82	1.53	0.063	0.67
	10	6.5	151.6	5.5	35.8	4.89	1.95	0.065	1.00
	11	5.2	1168	3.7	37.9	3.46	1.38	0.073	0.00
	12	7.5	1323	3.6	37.9	3.81	1.52	0.080(k)	0.00
	13	7.5	1323	2.3	37.9	2.43	0.97	0.080(k)	1.01
	14	7.3	1422	5.3	48.6	2.66	1.06	0.042	0.00
	15	5.1	3526	2.3	48.6	2.86	1.14	0.107(k)	0.00
	16	5.0	3669	1.9	48.6	2.46	0.98	0.107(k)	0.73

54

Table 2 (Continued)

Table Notes:
(a) Test identification according to references.
(b) f = sinusoidal input test frequency.
(c) $M/g_r = 386EI/(4f^2 L^2)$
  where M = moment at center of pipe span, in.–lb
    g_r = response acceleration
    E = modulus of elasticity, 30,000 ksi used
    I = section modulus of pipe cross section, in.
    f = sinusoidal input test frequency, Hz
    L = pipe span length, in.
(d) g_{mi} = maximum input acceleration during each test (from figures in the references).
(e) Sy = yield strength of pipe material (from references).
(f) $S = (M/g_{gr})g_{mi}/(2\zeta Z)$
  where ζ = damping factor, 0.02 or 0.05
    Z = section modulus of pipe cross section, in.
(g) Sw = stress at center of pipe span due to weight.
(h) Sp = stress due to internal pressure = $PD/(2t)$
  where P = internal pressure
    D = pipe mean diameter = $D_o - t$
    t = pipe wall thickness
    (See Table 1 for values of P, D_o, and t.)
(i) Straight pipe test of Reference 7.
(j) No g_{mi} given in Reference 9 for Test 2.
(k) In Reference 10, Tests 12, 13, 15, and 16, the pipe was filled with water.

Table 3 Reference 11 Pipe Tests: Limit Moments and Measured Moments

Test No.	Run No.	Type (a)	Pipe NPS	Pipe Sch.	Pipe Mtl. (b)	Sy, ksi (c)	$\frac{PD_o}{2tSy}$ (d)	M_L (e)	M_m (f)	M_L/M_m
9	6	T	6	40	SS	40.8	0.472	420	540	1.29
10	7	T	6	40	SS	40.8	0.278	446	491	1.10
11	6	T	6	10	SS	39.7	0.244	219	143	0.65
12	6	T	6	40	SS	40.8	0.472	420	492	1.17
14	6	T	6	40	CS	41.5	0.464	429	564	1.32
15	10	R	4	40	SS	37.0	0.413	148	189	1.27
16	6	R	4	40	CS	49.5	0.309	205	260	1.27
33	--	P	6	40	CS	44.5	0.255	490	532	1.09
34	12	P	6	40	CS	44.5	0.255	490	605	1.23
40	5	R	4	40	SS	37.0	0.000	159	202	1.27

(a)   T = 6x6x6 ANSI B16.9 tee, fixed at both run ends, branch loaded.
      R = 4 NPS pipe between 8x4 and 6x4 ANSI B16.9 reducers.
      P = straight pipe.
      Maximum loads are due to earthquake-type dynamic input, except for Test 33, during which sinesweep dynamic input was used.
(b)   SS = stainless steel, SA312 Type 316; CS = carbon steel, SA106–B.
(c) Sy = yield strength of material, ksi (from Appendix D of Reference 11).
      For tees (no data for pipe), tee data were used.
      For reducers, pipe data were used.
      For pipe, Sch. 40 pipe data were used.
(d) P = internal pressure; D = mean diameter of pipe; t = nominal wall thickness of pipe.
(e) M_L = calculated limit moment, in.–kip, = $D^2 tSy[1-0.75(PD/2tSy)^2]^{1/2}$
(f) M_m = maximum measured dynamic moment, in.–kip (from Appendix B of Reference 11).
      For Tests 12 and 14, M_m was adjusted by dividing the Reference 11 measured moment by 1.09 to obtain estimate of measured moment at the failure location.

Table 4 Reference 11 Elbow (6 NPS, 90°) Tests: Limit Moments and Measured Moments

Test/ Run	Elbow Sch. (a)	Elbow Mtl. (b)	Elbow h (c)	Sy, ksi (d)	Test Plane	$PD_o/2tSy$	M_L (e)	M_m (f)	M_L/M_m
1/8	80	CS	0.41	40.0	In	0.269	424	569	1.34
2/8	80	CS	0.41	40.0	Out	0.269	?	574	?
3/11	10	SS	0.11	34.0	In	0.285	52.3	163	3.12
4/?	40	CS	0.25	47.8	In	0.237	246	396	1.61
5/8	40	CS	0.25	47.8	In	0.403	246	478	1.94
6/8	40	SS	0.25	54.2	In	0.355	279	457	1.64
7/8	40	SS	0.25	54.2	In	0.209	279	426	1.53
8/8	40	SS	0.25	54.2	In	0.000	232	342	1.47
13/10	40	CS	0.17	47.0	In	0.241	189	400	2.12
17/?	40	CS	0.17	47.0	Tor	0.241	?	----	---
19/8	40	SS	0.25	54.0	In	0.525	278	450	1.62
23/4	40	CS	0.25	42.3	In	0.268	217	470	2.16(g)
25/14	10	SS	0.11	34.0	In	0.570	52.3	380?	7.31(h)
26/?	40	CS	0.25	42.3	In	0.455	217	----	----
30/4	10	SS	0.11	34.0	In	0.285	52.3	112	2.14
31/7	10	SS	0.11	38.6	In	0.251	59.4	150	2.53
35/?	40	CS	0.25	42.3	In	0.455	217	394	1.82
37/5	10	SS	0.11	34.0	In	0.000	43.6	57	1.31
41/?	40	CS	0.25	44.0	In	0.438	226	398	1.76

(a) All except Tests 13 and 17, 9-in. bend radius; Tests 13 and 17, 6-in. bend radius.
Maximum loads are due to earthquake-type dynamic input, except for Test 25, during which dynamic input in the middle-range frequency was used, and Test 26, during which sinesweep dynamic input was used.
(b) CS = carbon steel, SA106–B; SS = stainless steel, SA312 Type 316.
(c) h = elbow parameter = tR/r^2
where t = elbow nominal wall thickness
R = elbow bend radius
r = mean elbow cross-section radius
(d) Sy = material yield strength (from Appendix D of Reference 11).
(e) M_L = limit moment calculated using Equation (9), in.-kip; conceptually, in-plane, closing limit moment.
(f) M_m = maximum measured moment, in.-kip (from Appendix B of Reference 11).
(g) Assembly restrained with a strut. Significance of the measured moment is not clear.
(h) Reference 11, Appendix B, states: "Mid-freq. moment measuring method is still in study, results will be changed."

NUREG–1367

Table 5 Reference 3 Static and Dynamic In–Plane Moment Capacity Tests on Elbows
(See Figure 12 for plot of these data.)

Test Iden. (a)	Mtl. (b)	Sy, ksi (c)	$PD_0/2t$, ksi (d)	M, in.–kip (e)	D^2tSy, in.–kip	M/D^2tSy	h (f)
22(2)	CS	50.0	0	261+	563.6	0.46+	0.25
22(5)	CS	50.0	17.0	347+	563.6	0.61+	0.25
22(8)	CS	37.8	0	450+	626.3	0.72+	0.41
22(11)	CS	39.6	0	202+	446.4	0.45+	0.17
22(15)	SS	37.7	0	206+	425.0	0.48+	0.25
22(16)	SS	37.7	0	202+	425.0	0.48+	0.25
22(17)	SS	35.6	0	200+	401.3	0.50+	0.17
22(18)	SS	35.4	0	381	586.5	0.65	0.41
22(19)	CS	46.0	0	202	518.5	0.39	0.17
22(20)	CS	34.6	0	369	573.3	0.64	0.27
23(1)	SS	36.3	0	3300	6515	0.51	0.15
13(1)	CS	(45)	15.0	269+	488.1	0.55+	0.26
13(5)	SS	(35)	0	166+	379.6	0.44+	0.26
13(6)	SS	(35)	18.6	313	379.6	0.82	0.26
13(7)	SS	(35)	0	122	274.3	0.44	0.18
13(8)	SS	(35)	0	79	221.4	0.36	0.14
13(9)	SS	(35)	0	78+	106.4	0.73+	0.40
24(4)	SS	(35)	16.5	+/–43.8(g)	47.5	0.92	0.18
24(5)	SS	(35)	17.3	+/–45.0(g)	47.5	0.95	0.18

(a) Identification according to Table 4 of Reference 3.
(b) CS = carbon steel; SS = stainless steel.
(c) Sy = yield strength as listed in Reference 3. For References 13 and 24 in Reference 3, yield strengths were not given. Typical values of 45 ksi for carbon steel and 35 ksi for stainless steel were used.
(d) P = internal pressure; D = mean diameter of elbow; t = nominal wall thickness of elbow.
(e) From Table 4 of Reference 3, column headed "M_m". A "+" indicates that the moment capacity was not reached in the static loading test.
(f) h = elbow parameter = tR/r^2
where t = elbow wall thickness
       R = elbow bend radius
       r = elbow mean cross–section radius
(g) These values were derived from sinusoidal dynamic loading tests.

Table 6 Reference 11 Pipe Tests: Comparisons with 2Sy Limit

Test/ Run	Type (a)	Pipe			Sy, ksi (c)	2% Damping		5% Damping		Sw/Sy (f)	Sp/Sy (g)
		NPS	Sch.	Mtl. (b)		S, ksi (d)	S/2Sy	S, ksi (e)	S/2Sy		
9/6	T	6	40	SS	40.8	589	7.2	330	4.0	0.02	0.47
10/7	T	6	40	SS	40.8	600	7.4	335	4.1	0.02	0.28
11/6	T	6	10	SS	39.7	269	3.4	178	2.2	0.04	0.24
12/6	T	6	40	SS	40.8	737	9.0	401	4.9	0.02	0.47
14/6	T	6	40	CS	41.5	542	6.5	304	3.7	0.02	0.46
15/9	R	4	40	SS	37.0	787	11	428	5.8	0.06	0.41
16/6	R	4	40	CS	49.5	1979	20	1011	10	0.04	0.31
33/?	P	6	40	CS	44.5	---	---	---	---	0.00	0.25
34/12	P	6	40	CS	44.5	731	8.2	419	4.7	0.01	0.25
40/5	R	4	40	SS	37.0	1345	18	786	11	0.06	0.00

(a) T = 6x6x6 ANSI B16.9 tee, fixed at both run ends, branch loaded..
R = 4 NPS pipe between 8x4 and 6x4 ANSI B16.9 reducers.
P = straight pipe.
Maximum loads are due to earthquake-type dynamic input, except for Test 33, during which sinesweep dynamic input was used.

(b) SS = stainless steel, SA312 Type 316; CS = carbon steel, SA106–B.

(c) Sy = material yield strength (from Appendix D of Reference 11).
For tees (no data for pipe), tee data were used.
For reducers, pipe data were used.
For pipe, Sch. 40 pipe data were used.

(d) From Appendix B of Reference 11, Case 2, B_2M/Z with B_2 = 1.00. Appendix B states: "Case 2 Actual tested time history used 2% damping amplified response spectrum + /–15% broadening response spectrum analysis."

(e) From Appendix B of Reference 11, Case 3, B_2M/Z with B_2 = 1.00. Appendix B states: "Case 3 Same as Case 2 except using 5% damping."

(f) Sw = stress due to weight = M_w/Z where M_w = moment due to weight.

(g) Sp = stress due to internal pressure = $PD/2t$
where P = internal pressure
D = pipe mean diameter
t = pipe nominal wall thickness

NUREG-1367

Test/ Run	Elbow Sch. (a)	Elbow Mtl. (b)	Elbow Sy, ksi (c)	Test Plane	2% Damping S, ksi (d)	2% Damping S/2Sy	5% Damping S, ksi (e)	5% Damping S/2Sy	Sw/Sy (f)	Sp/Sy (g)
1/8	80	CS	40.0	In	890	11	547	6.8	0.01	0.27
2/8	80	CS	40.0	Out	897	11	501	6.3	0.01	0.27
3/10	10	SS	34.0	In	1276	19	752	11	0.04	0.28
4/?	40	CS	47.8	In	1057	11	648	6.8	0.01	0.24
5/8	40	CS	47.8	In	1238	13	674	7.0	0.01	0.40
6/8	40	SS	54.2	In	1158	11	634	5.8	0.01	0.36
7/8	40	SS	54.2	In	1392	13	756	7.0	0.01	0.21
8/8	40	SS	54.2	In	1442	13	776	7.2	0.01	0.00
13/10	40	CS	47.0	In	1255	13	679	7.2	0.02	0.24
17/?	40	CS	47.0	Tor	----	----	----	----	0.02	0.24
19/8	40	SS	54.0	In	1331	12	707	6.5	0.01	0.52
25/15	10	SS	34.0	In	1628	24	990	15	0.04	0.57
26/?	40	CS	42.3	In	----	----	----	----	0.01	0.46
30/4	10	SS	34.0	In	620	9.1	356	5.2	0.32	0.28
31/11	10	SS	38.6	In	1391	18	738	9.6	0.04	0.25
35/?	40	CS	42.3	In	----	----	----	----	0.08	0.46
37/5	10	SS	34.0	In	651	9.6	375	5.5	0.32	0.00
41/?	40	CS	44.0	In	----	----	----	----	0.01	0.44

(a) All except Tests 13 and 17, 9-in. bend radius; Tests 13 and 17, 6-in. bend radius.
　　Maximum loads are due to earthquake-type dynamic input, except for Test 25, during which dynamic input in the middle-range frequency was used, and Test 26, during which sinesweep dynamic input was used.
(b) Elbow material: CS = carbon steel, SA106–B; SS = stainless steel, SA312 Type 316.
(c) Sy = material yield strength (from Appendix D of Reference 11).
(d) From Appendix B of Reference 11.
　　Response spectrum analysis based on 2% damping, +/–15% peak broadening.
(e) Same as (d), except 5% damping was used.
(f) Sw = stress due to weight = $B_2 M_w / Z$ where M_w = moment due to weight at mid–arc of elbow.
(g) Sp = stress due to internal pressure = $PD/2t$
　　where P = internal pressure
　　　　　D = elbow mean diameter
　　　　　t = elbow nominal wall thickness

Table 8 Reference 11 Tests on Other Components: Comparisons with 2Sy Limit

Test/ Run	Type (a)	Pipe Size	Pipe Sch.	Pipe Mtl. (b)	Pipe Sy, ksi (c)	2% Damping S, ksi (d)	2% Damping $S/2Sy$	5% Damping S, ksi (e)	5% Damping $S/2Sy$	Sw/Sy (f)	Sp/Sy (g)
18/6	RFT	(a)	(a)	CS	53.4	770	7.2	456	4.3	0.08	0.24
20/7	NZ	(a)	(a)	SS	48.7	770	7.9	436	4.5	0.05	0.34
36/8	TR	6	40	CS	45.5	902	9.9	610	6.7	0.05	0.42
38/6	TB	6	40	SS	40.1	1185	15	654	8.2	0.05	0.48
39/4	TB	6	40	SS	40.1	1248	16	674	8.4	0.04	0.00

(a) RFT = reinforced (with pad) fabricated tee; 4 NPS, Sch. 40 branch; 8 NPS, Sch. 40 run; pad thickness = 0.322 in.
     NZ = nozzle (see Figure 13).
     TR = 6x6x6 ANSI B16.9 tee, loaded through run.
     TB = 6x6x6 ANSI B16.9 tee, fixed at one run end, branch loaded.
(b) Pipe material: CS = carbon steel, SA106–B; SS = stainless steel, SA312 Type 316.
(c) Sy = material yield strength (from Appendix D of Reference 11). For Tests 18 and 20, run pipe material yield strengths.
(d) Moments from Appendix B of Reference 11.
     Response spectrum analysis based on 2% damping, +/–15% peak broadening.
     See text for conversion of moments to stresses.
(e) Same as (d), except 5% damping was used.
(f) Sw = stress due to weight = $B_2 M_w/Z$ where M_w = moment due to weight.
(g) Sp = stress due to internal pressure = $PD/2t$
     where P = internal pressure
        D = run pipe mean diameter
        t = run pipe nominal wall thickness

Table 9 Reference 13 and Other Piping System Tests: Comparison with Elastic Analyses

Ref.	Test Location (a)	System Description	Mtl. (b)	$S/2S_y$ (c)	Damping, % (c)
13	ETEC	System 1, 6 NPS and 3 NPS, Sch. 40 See Figure 15 and Section 5	CS	16	5
13	ETEC	System 2, 6 NPS and 4 NPS, Sch. 40 See Figure 16 and Section 5	SS	12	5
14	HEDL	1 NPS, Sch. 40 See Figure 20	SS	12	5
15	ANCO	Z bend, 4 NPS, Sch. 40	CS	1.6	2
16	ANCO	8 NPS and 6 NPS, Sch. 40 No branches	CS	3	3
16	ANCO	8 NPS and 6 NPS, Sch. 40 Two 3 NPS, Sch. 40 branches	CS	----	----
17	ETEC	3 NPS, Sch. 40 One 3 NPS, Sch. 40 branch	CS	14	5
17	ETEC	6 NPS, Sch. 40 One 3 NPS, Sch. 40 branch	CS	8.3	5
18	KWU (Germany)	4.5-in. outer diameter, 0.165-in. wall thickness 2.38-in. outer diameter, 0.114-in. wall branch	SS	2.1	2
19	HDR (Germany)	18-in.- to 4.5-in.-outer-diameter pipes D/t = about 15	SS	2.4	3
20	Tadotsu (Japan)	1/2.5 scale model of one loop of PWR primary coolant system D/t = about 12	SS	(6.6)	(d)

(a) ETEC = Energy Technology Engineering Center, Canoga Park, California
     HEDL = Hanford Engineering Development Laboratory, Richland, Washington
     ANCO = ANCO Engineers, Culver City, California
     KWU = Kraftwerk Union, Aktiengesellschaft, Federal Republic of Germany
     HDR = Heissdampfreaktor, Kahl/Main, Federal Republic of Germany
     Tadotsu = Tadotsu Engineering Laboratory, Tadotsu-cho, Kagawa Prefecture, Japan
(b) CS = carbon steel pipe material (e.g., A106–B)
     SS = austenitic stainless steel pipe material (e.g., A312 Type 304)
(c) S = calculated stress using response spectrum analysis with indicated damping, except for Reference 20.
     S for Reference 20 is from a time history analysis.
     S_y = yield strength of piping material.
(d) Time history analysis

NRC FORM 335 (2-89) NRCM 1102, 3201, 3202	U.S. NUCLEAR REGULATORY COMMISSION **BIBLIOGRAPHIC DATA SHEET** (See instructions on the reverse)	1. REPORT NUMBER (Assigned by NRC, Add Vol., Supp., Rev., and Addendum Num- bers, if any.) NUREG–1367

2. TITLE AND SUBTITLE	3. DATE REPORT PUBLISHED
Functional Capability of Piping Systems	MONTH / YEAR November / 1992
	4. FIN OR GRANT NUMBER

5. AUTHOR(S) D. Terao, E. C. Rodabaugh	6. TYPE OF REPORT Technical
	7. PERIOD COVERED (Inclusive Dates)

8. PERFORMING ORGANIZATION – NAME AND ADDRESS (If NRC, provide Division, Office or Region, U.S. Nuclear Regulatory Commission, and mailing address; if contractor, provide name and mailing address.)

Division of Engineering
Office of Nuclear Reactor Regulation
U.S. Nuclear Regulatory Commission
Washington, DC 20555

9. SPONSORING ORGANIZATION – NAME AND ADDRESS (If NRC, type "Same as above"; if contractor, provide NRC Division, Office or Region, U.S. Nuclear Regulatory Commission, and mailing address.)

Same as above

10. SUPPLEMENTARY NOTES

11. ABSTRACT (200 words or less)

General Design Criterion 1 of Appendix A to Part 50 of Title 10 of the *Code of Federal Regulations* requires, in part, that structures, systems, and components important to safety be designed to withstand the effects of earthquakes without a loss of capability to perform their safety function. The function of a piping system is to convey fluids from one location to another. The functional capability of a piping system might be lost if, for example, the cross-sectional flow area of the pipe were deformed to such an extent that the required flow through the pipe would be restricted.

The objective of this report is to examine the present rules in the American Society of Mechanical Engineers Boiler and Pressure Vessel Code, Section III, and potential changes to these rules, to determine if they are adequate for ensuring the functional capability of safety-related piping systems in nuclear power plants.

12. KEY WORDS/DESCRIPTORS (List words or phrases that will assist researchers in locating the report.) piping operability limits functional capability stress limit	13. AVAILABILITY STATEMENT Unlimited
	14. SECURITY CLASSIFICATION (This Page) Unclassified (This Report) Unclassified
	15. NUMBER OF PAGES
	16. PRICE

NRC FORM 335 (2-89)

FUNCTIONAL CAPABILITY OF PIPING SYSTEMS

UNITED STATES
NUCLEAR REGULATORY COMMISSION
WASHINGTON, D.C. 20555-0001

OFFICIAL BUSINESS
PENALTY FOR PRIVATE USE, $300

FIRST CLASS MAIL
POSTAGE AND FEES PAID
USNRC
PERMIT NO. G-67

www.ingramcontent.com/pod-product-compliance
Lightning Source LLC
Chambersburg PA
CBHW081845170526
45167CB00007B/2906